TREATMENT PLANT HYDRAULICS FOR ENVIRONMENTAL ENGINEERS

Larry D. Benefield
Associate Professor
Auburn University

Joseph F. Judkins, Jr.
Engineer
Paul B. Krebs and Associates, Inc.

A. David Parr
Associate Professor
University of Kansas

PRENTICE-HALL, INC.   Englewood Cliffs, NJ 07632

Library of Congress Cataloging in Publication Data

Benefield, Larry D.
  Treatment plant hydraulics for environmental engineers.

  Includes index.
    1. Sewage—Purification—Activated sludge process.
  2. Sewage disposal plants—Design and construction.
  3. Hydraulics—Problems, exercises, etc.   I. Judkins,
  Joseph F.   II. Parr, A. David, 1946–   .   III. Title.
  TD756.B46 1984   628.3   83-3352
  ISBN 0-13-930248-4

Editorial/production supervision and interior design: *Shari Ingerman*
Cover design: *Marvin Warshaw*
Manufacturing buyer: *Tony Caruso*

Printed in the United States of America

10 9 8 7 6 5 4 3 2 1

ISBN 0-13-930248-4

Prentice-Hall International, Inc., *London*
Prentice-Hall of Australia Pty. Limited, *Sydney*
Editora Prentice-Hall do Brasil, Ltda., *Rio de Janeiro*
Prentice-Hall Canada Inc., *Toronto*
Prentice-Hall of India Private Limited, *New Delhi*
Prentice-Hall of Japan, Inc., *Tokyo*
Prentice-Hall of Southeast Asia Pte. Ltd., *Singapore*
Whitehall Books Limited, *Wellington, New Zealand*

CONTENTS

PREFACE

The design of water and wastewater treatment plants involves both *process design* and *hydraulic design*. Each of these segments of design must be performed correctly if the treatment plant is to function properly and provide the desired degree of treatment. Most engineering courses are structured to emphasize process mechanisms and control variables and thus prepare students to perform *process design*. In addition, many textbooks are available to which engineers can turn for information on this subject.

Unfortunately, the *hydraulic design* of water and wastewater treatment plants is largely ignored in most engineering curricula. Undergraduate courses in fluid mechanics and hydraulics emphasize fundamentals and introduce problems on pipe flow, pumps, and open channel flow. Environmental engineering courses may address the hydraulics of a particular process, such as a rapid sand filter, but usually do not consider overall plant hydraulics. Consequently, many graduates are not familiar with the hydraulic design of treatment plants and, since hydraulic design is normally not included in engineering course work, textbooks specifically intended to provide guidance on the subject are not currently available.

The authors feel that a need exists for a book that brings together the information required for the *hydraulic design* of water and wastewater treatment facilities. This book was written in an attempt to help satisfy that need. The reader is assumed to have at least an elementary background in the principles of fluid mechanics. It is also assumed that the reader has an acquaintance with the utilization of digital computers to solve scientific problems.

The first five chapters provide a review of hydraulic fundamentals, emphasizing components and situations commonly encountered in treatment plant design. Chapter 6 presents a step-by-step example of the hydraulic design of an activated sludge treatment plant. The example does not attempt to optimize plant hydraulics and, in fact, an effort is made to introduce a variety of components and conditions that may not be found in a typical treatment plant. The example is intended to illustrate the importance of hydraulic control points in plant design and to show the reader the manner in which various units must operate compatibly to provide the desired flow profile. Once these concepts are understood, the reader should be able to adapt them to any plant configuration or processes dictated by a particular situation.

A word of appreciation is due to Joy Woodham and Dawn Horne, who typed the manuscript for publication, and to Mary Benefield for editing the final version of the manuscript.

Larry D. Benefield
Joseph F. Judkins, Jr.
A. David Parr

1

FLOW IN PIPES

The purpose of this chapter is to review briefly the hydraulic fundamentals required to solve problems related to the flow of water in pipes. It is appropriate to begin with a discussion of the types of flow a design engineer is likely to encounter. The two fundamental types of fluid flow are known as laminar and turbulent. *Laminar flow* is characterized by fluid particles that move in straight lines or parallel layers, whereas *turbulent flow* is characterized by random movement of fluid particles (see Fig. 1-1). According to Brater and King (1976), the greater energy loss in turbulent flow is probably the most important practical difference between laminar and turbulent flow.

On the basis of discharge, a flow may be classified as steady or unsteady. In *steady flow,* the discharge and depth at a particular cross section do not vary with time. The flow is *unsteady* when the discharge or depth at a particular point varies with time. A steady or unsteady flow may be described as spatially variable, uniform, or non-uniform. *Spatially variable flow* (a subclassification of nonuniform flow) occurs when the discharge varies along a specified reach or length of channel. *Uniform flow* occurs when the cross-sectional area of the fluid remains constant along a specified reach of channel, while *nonuniform flow* arises when the cross-sectional area of the fluid varies along a specified length of channel (see Fig. 1-2).

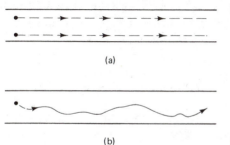

(a)

(b)   **FIGURE 1-1** *(a) Laminar flow; (b) turbulent flow.*

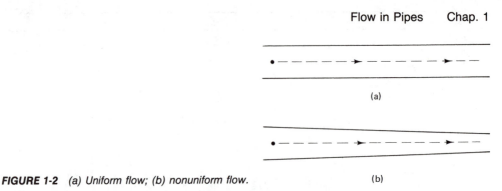

FIGURE 1-2 (a) Uniform flow; (b) nonuniform flow.   (b)

1.1 CONSERVATION LAWS

The conservation laws of mass, momentum, and energy are the three fundamental concepts used in the solution of problems related to fluid flow. The simplified equations derived from the conservation principles of mass, energy, and momentum are commonly called the continuity equation, the Bernoulli equation, and the linear momentum equation, respectively.

Conservation of Mass

According to the Law of Conservation of Mass, material is neither created nor destroyed. Hence, any mass of material that enters a system must either accumulate in the system or leave the system. This fundamental statement is expressed by Eq. (1-1).

$$\begin{bmatrix} \text{Accumulation} \\ \text{of mass in} \\ \text{the system} \end{bmatrix} = \begin{bmatrix} \text{total mass of} \\ \text{material that has} \\ \text{entered system} \end{bmatrix} - \begin{bmatrix} \text{total mass of} \\ \text{material that has} \\ \text{entered system} \end{bmatrix} \qquad \textbf{(1-1)}$$

The majority of the systems encountered in hydraulic design are continuous flow systems. For this case Eq. (1-1) has the form

$$\begin{bmatrix} \text{Rate of accumulation} \\ \text{of mass in the} \\ \text{system} \end{bmatrix} = \begin{bmatrix} \text{total rate of} \\ \text{mass flow} \\ \text{into system} \end{bmatrix} - \begin{bmatrix} \text{total rate of} \\ \text{mass flow out} \\ \text{of the system} \end{bmatrix} \qquad \textbf{(1-2)}$$

This situation is illustrated schematically in Fig. 1-3. In this figure Q represents the volumetric flow rate (i.e., volume of fluid flowing per unit time). A mathematical relationship that describes the situation presented in Fig. 1-3 is

$$\rho_{\text{basin}} \left[\frac{\Delta V}{\Delta t} \right] = \rho_{\text{in}} Q_{\text{in}} - \rho_{\text{out}} Q_{\text{out}} \qquad \textbf{(1-3)}$$

where ρ represents the mean fluid density and V represents volume within control.

In a case where there is no storage (such as a pipe flowing full) Eq. (1-3) reduces

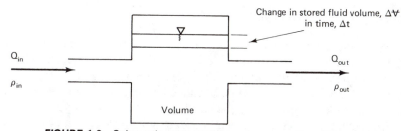

FIGURE 1-3   *Schematic representation of conservation of mass.*

to the form

$$0 = \rho_{in}Q_{in} - \rho_{out}Q_{out} \tag{1-4}$$

or for incompressible fluids

$$Q_{in} = Q_{out} \tag{1-5}$$

Note: ρ may not be constant even for incompressible fluids. However, most waste-waters are considered to have a constant density.

The volumetric flow rate may be expressed in terms of velocity and area as

$$Q = AV \tag{1-6}$$

where Q = volumetric flow rate, length^3 time^{-1}
$\quad\quad\quad A$ = cross-sectional area of flow, length^2
$\quad\quad\quad V$ = average velocity of the flow through the section, length time^{-1}

Substituting for Q in Eq. (1-5) from Eq. (1-6) gives

$$(AV)_{in} = (AV)_{out} \tag{1-7}$$

EXAMPLE PROBLEM 1-1:   A 4-in. pipe is connected to a 6-in. pipe. If the average velocity of flow in the 6-in. pipe is 20 ft/sec (fps), what is the average velocity of flow in the 4-in. pipe?

Solution: Apply Eq. (1-7) and solve for velocity in the 4-in. section.

$$V_4 = \frac{A_6 V_6}{A_4}$$

$$= \frac{[\pi(6/12)^2/4](20)}{[\pi(4/12)^2/4]}$$

$$V_4 = \textbf{45 fps}$$

Conservation of Energy

In most hydraulic problems encountered by environmental engineers, two forms of energy are important. These are kinetic and potential energy. The kinetic energy of a mass, m, moving with a velocity, V, is given by $mV^2/2$. Two types of potential energy are of interest. The first type is related to the height of the mass above an arbitrary datum (elevation), Z, and the acceleration due to gravity, g, and is given by mgZ. The second type of potential energy is due to the pressure, p, of the flowing fluid and is given by pm/ρ. These different types of energy may be summed to give an expression for total energy.

$$E_T = \frac{mV^2}{2} + mgZ + \frac{pm}{\rho} \qquad \text{(1-8)}$$

Equation (1-8) is more useful when expressed on a total energy per unit mass of fluid basis. This can be accomplished by dividing Eq. (1-8) by mass of fluid.

$$\frac{E_T}{m} = \frac{V^2}{2} + gZ + \frac{p}{\rho} \qquad \text{(1-9)}$$

This equation assumes one-dimensional flow where the velocity is constant at a cross section. When nonuniform velocity profiles are considered, the velocity head term must be multiplied by the kinetic energy correction factor.

The Law of Conservation of Energy is a statement of the First Law of Thermodynamics, which says that energy cannot be created or destroyed but can be transformed from one form to another. Consider the pipe section presented in Fig. 1-4. As fluid flows between sections 1 and 2, fluid friction will convert some of the useful energy into heat energy. Hence, when writing a flow energy balance between sections 1 and 2, the energy loss due to friction must be accounted for.

$$\frac{(E_T)_1}{m} = \frac{(E_T)_2}{m} + \frac{(E_T)_L}{m} \qquad \text{(1-10)}$$

where $(E_T)_L/m$ represents the useful energy loss due to friction per unit mass of fluid. Substituting for the energy terms in Eq. (1-10) from Eq. (1-9) gives

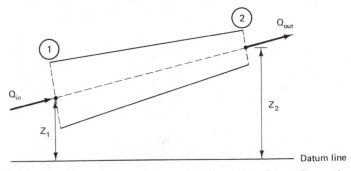

FIGURE 1-4 *Flow section for the development of the Bernoulli equation.*

$$\frac{V_1^2}{2} + gZ_1 + \frac{p_1}{\rho} = \frac{V_2^2}{2} + gZ_2 + \frac{p_2}{\rho} + \frac{(E_T)_L}{m} \qquad \textbf{(1-11)}$$

If mechanical energy is added to or taken from the fluid over the section of interest, this must be accounted for in the energy balance equation. In this regard, a pump adds mechanical energy to the system, while a turbine extracts mechanical energy from the system.

The Bernoulli equation is obtained by dividing each term in Eq. (1-11) by the acceleration of gravity so that each term will have the dimension of length

$$\frac{V_1^2}{2g} + Z_1 + \frac{p_1}{\gamma} = \frac{V_2^2}{2g} + Z_2 + \frac{p_2}{\gamma} + h_L \qquad \textbf{(1-12)}$$

where γ represents the specific weight of the fluid (i.e., $g\rho$) and h_L represents the energy loss per unit mass of fluid due to friction. Because each term in Eq. (1-12) has the units of length, each term is referred to as a type of head. *Velocity head* is the designation given to the $V^2/2g$ term, whereas the Z term is called *elevation head,* the p/γ term is called *pressure head,* and the h_L term is called *head loss.*

The sum of the velocity head, the elevation head, and the pressure head is referred to as the *total head,* whereas the sum of only the pressure head and the elevation head is called the *piezometric head.* Piezometric head represents the height to which fluid would rise in a pipe with one of its ends inserted into the flow field perpendicular to the direction of flow. The hydraulic grade line (HGL) is a line that shows how the piezometric head varies over a particular reach of the system of interest, whereas the energy grade line (EGL) indicates the variation in the total head over the reach of interest. The hydraulic grade line and the energy grade line are illustrated for a particular flow system in Fig. 1-5. The difference in elevation between the EGL and the HGL is the velocity head, $V^2/2g$.

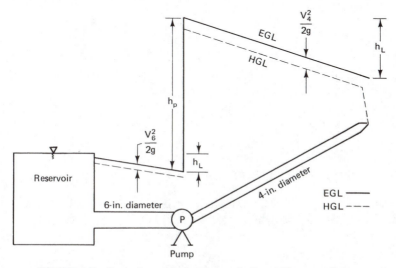

FIGURE 1-5   *Illustration of hydraulic grade line and energy line.*

Equation (1-12) may also be applied to open channel flow problems, so long as points 1 and 2 are taken along the same streamline (e.g. along the water surface).

EXAMPLE PROBLEM 1-2: Calculate the head loss due to friction and other factors in the piping system shown below. The pipe is 12 in. in diameter and 500 ft long and passes a flow of 40 cfs. The water discharges as a free jet at a point 2.

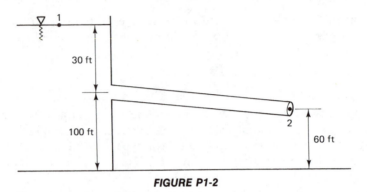

FIGURE P1-2

Solution:

1. Write the Bernoulli equation between points 1 and 2.

$$\frac{V_1^2}{2g} + \frac{p_1}{\gamma} + Z_1 = \frac{V_2^2}{2g} + \frac{p_2}{\gamma} + Z_2 + h_L$$

2. Evaluate each term in the Bernoulli equation and solve for h_L. Since both points 1 and 2 are at atmospheric pressure, the pressure head at both points is zero. Neglecting the fluid velocity in the reservoir and solving the Bernoulli equation for h_L gives

$$h_L = (Z_1 - Z_2) - \frac{V_2^2}{2g}$$

$$h_L = 70 - \frac{Q^2}{2gA^2}$$

$$= 70 - \frac{(40)^2}{64.4(\pi)^2 (0.5)^4}$$

$$h_L = \textbf{29.7 ft}$$

Conservation of Momentum

Momentum is defined as the product of mass and velocity. Hence, for an incompressible fluid, the rate at which momentum is carried across a section is defined mathematically as

$$\overline{M} = \rho Q V \qquad\qquad (1\text{-}13)$$

where $\overline{M}$ = momentum flux at the section, mass length time^{-2}

ρ = density of fluid, mass length^{-3}

Q = volumetric flow rate, length^3 time^{-1}

V = average velocity at the section, length time^{-1}

Substituting for Q in Eq. (1-13) from Eq. (1-6) produces an alternate form of Eq. (1-13)

$$\overline{M} = \rho A V^2 \qquad \text{(1-14)}$$

where A represents the cross-sectional area of flow.

The Law of Conservation of Momentum states that the sum of the external forces acting on a fluid system equals the rate of change of momentum of the system. For example, consider steady flow through the pipe bend shown in Fig. 1-6. The momentum carried across areas A_1 and A_2 in time Δt are given, respectively, by

$$M_1 = (\rho Q \Delta t)V_1 = (\rho A_1 V_1 \Delta t)V_1 \qquad \text{(1-15)}$$

$$M_2 = (\rho Q \Delta t)V_2 = (\rho A_2 V_2 \Delta t)V_2 \qquad \text{(1-16)}$$

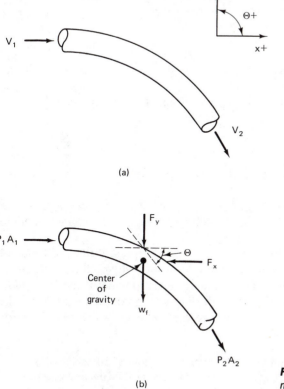

(a)

(b)

FIGURE 1-6 *Pipe bend to illustrate momentum principle.*

The change in momentum between A_1 and A_2 is, therefore,

$$\Delta M = (\rho A_2 V_2 \Delta t)V_2 - (\rho A_1 V_1 \Delta t)V_1$$

The net force acting on the fluid between A_1 and A_2 is, thus, equal to the *rate of change of momentum* between A_1 and A_2. Hence

$$\Sigma F = \frac{\Delta M}{\Delta t}$$

or

$$\Sigma F = \frac{(\rho Q \Delta t)V_2 - (\rho Q \Delta t)V_1}{\Delta t} \qquad (1\text{-}17)$$

where $\quad \Sigma F =$ net force acting on the fluid between A_1 and A_2

Equation (1-17) is a vector equation. Considering the free body diagram shown in Fig. 1-6(b), it is also possible to write the component equation of Eq. (1-17) in the x and y directions as follows:

$$\Sigma F_x = \begin{bmatrix} \text{net force acting on the} \\ \text{fluid in the } x \text{ direction} \end{bmatrix} = \begin{bmatrix} \text{rate of change in momentum} \\ \text{in the } x \text{ direction} \end{bmatrix}$$

or

$$p_1 A_1 - F_x - p_2 A_2 \cos \Theta = \rho Q (V_2 \cos \Theta - V_1) \qquad (1\text{-}18)$$

$$\Sigma F_y = \begin{bmatrix} \text{net force acting on the} \\ \text{fluid in the } y \text{ direction} \end{bmatrix} = \begin{bmatrix} \text{rate of change in momentum} \\ \text{in the } y \text{ direction} \end{bmatrix}$$

or

$$p_2 A_2 \sin \Theta - W_F - F_y = -\rho Q V_2 \sin \Theta \qquad (1\text{-}19)$$

where W_F is the weight of the fluid in the bend section. The negative sign on the right-hand side of Eq. (1-19) arises because the velocity component at point 2 is in the negative y direction.

For the two-dimensional case, Eqs. (1-18) and (1-19) represent the two forms of the *momentum equation*. This equation is important in many hydraulic problems. It is often employed in conjunction with the continuity equation and many times additionally with the Bernoulli equation. One of the most common applications of the momentum equation is to solve problems where a *change in velocity* or *direction* occurs.

EXAMPLE PROBLEM 1-3: An 8-in. pipeline carries a flow of 10 cubic feet per second (cfs). Compute the magnitude of the force exerted by the fluid on the pipe when the flow passes through a 90° bend. Assume the pipe is horizontal.

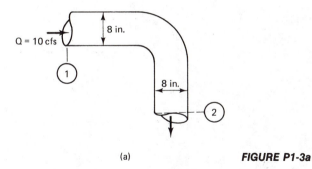

(a)   *FIGURE P1-3a*

The pipe discharges into the atmosphere at point 2. Assume the head loss through the bend is given by $h_L = 0.21\ V_1^2/2g$.

Solution:

1. Construct the free body diagram of the flow section.

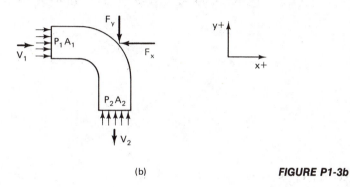

(b)   *FIGURE P1-3b*

2. Compute the average velocity of flow through the pipeline.

$$V_1 = V_2 = \frac{Q}{A}$$

$$= \frac{3}{[\pi(8/12)^2/4]}$$

$$= \textbf{8.6 fps}$$

3. Write Bernoulli's equation between sections 1 and 2 and evaluate the pressure at each section.

$$\frac{V_1^2}{2g} + Z_1 + \frac{p_1}{\gamma} = \frac{V_2^2}{2g} + Z_2 + \frac{p_2}{\gamma} + h_L$$

Since the pipe is horizontal, $Z_1 = Z_2$. Because the pipe discharges into the atmosphere at point 2, $p_2 = 0$. Thus,

$$p_1 = \gamma\left[\frac{V_2^2}{2g} - \frac{V_1^2}{2g} + h_L\right] = \gamma h_L = \frac{\gamma 0.21(8.6)^2}{2g} = \frac{(62.4)(0.21)(8.6)^2}{2(32.2)}$$
$$= \mathbf{15\ lb/ft^2}$$

4. Write the momentum equation for the flow section.

$$\Sigma F_x = p_1A_1 - F_x - p_2A_2\cos\Theta = \rho Q(V_2\cos\Theta - V_1)$$
$$= p_1A_1 - F_x - p_2A_2(0) = \rho Q(V_2(0) - V_1)$$
$$= p_1A_1 - F_x = -\rho QV_1$$
$$\Sigma F_y = p_2A_2\sin\Theta - F_y = -\rho QV_2\sin\Theta$$
$$= p_2A_2(1) - F_y = -\rho QV_2(1)$$
$$= p_2A_2 - F_y = -\rho QV_2$$

5. Compute the magnitude of the force acting on the bend.

$$F_R = \sqrt{F_x^2 + F_y^2}$$

(a) Compute the x component of the force.

$$F_x = p_1A_1 + \rho QV_1$$
$$= \left[15\frac{lb}{ft^2} \times 0.35\ ft^2\right] + \left[\left(\frac{62.4\ lb/ft^3}{32.2\ ft/s^2}\right)\left(10\ ft^2/s\right)\left(8.6\ ft/s\right)\right]$$
$$F_x = \mathbf{171.9\ lb}$$

The positive sign indicates that the assumed direction for the force F_x on the fluid was correct.

(b) Compute the y component of the force.

$$F_y = p_2A_2 + \rho QV_2$$
$$= 0 + \left[\left(\frac{62.4\ lb/ft^3}{32.2\ ft/s^2}\right)\left(10\ ft^3/s\right)\left(8.6\ ft/s\right)\right]$$
$$= \mathbf{166.6\ lb}$$

Therefore, the assumed direction of the force the bend exerts on the water was correct.

(c) Determine the resultant force on the bend. The forces that the water exerts on the bend are equal and opposite to the forces on the water, F_x

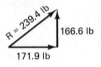

FIGURE P1-3c   (c)

and F_y. Therefore, the magnitude of the force on the bend is

$$F_B = \sqrt{(171.9)^2 + (166.6)^2}$$
$$= \mathbf{239.4\ lb}$$

Note: The momentum equation gives no information regarding the location of the resultant. This information is obtained by applying the angular momentum equation. This will not be discussed.

1.2 HEAD LOSSES DUE TO FRICTION

The loss of energy due to wall friction in pipe flow systems can be calculated with the Darcy–Weisbach equation

$$h_L = \frac{fLV^2}{2gD} = \frac{8fLQ^2}{\pi^2 gD^5} \qquad \textbf{(1-20)}$$

where h_L = energy loss, ft

  L = length of flow section of interest, ft

  D = pipe diameter, ft

  g = gravity constant, ft/s^2

  V = average velocity of flow, ft/s

  f = dimensionless friction factor

  Q = discharge, ft^3/s

The friction factor is a function of the Reynolds number, Re, and the relative roughness of the pipe wall, e/D. The e term represents the roughness of the pipe wall and is called the *equivalent roughness*. Although e cannot be measured directly, it can be determined from experimental measurements (see Table 1-1 for typical values of e for various materials).

Nikuradse performed a series of experiments in which he coated the inside of pipes with uniform-sized sand grains. By observing the effects produced by different grain sizes, he was able to illustrate graphically the relationship between f, Re, and e/D. The Reynolds number is given by the relationship

$$\text{Re} = \frac{VD}{\nu} = \frac{4Q}{\pi D\nu} \qquad \textbf{(1-21)}$$

where Q = discharge, ft^3/s

  V = average velocity of fluid, ft/s

  D = diameter of pipe, ft

  ν = kinematic viscosity of the fluid, ft^2/s

TABLE 1-1 *Values of equivalent roughness e for new commercial pipes after Jeppson (1977).*

	e	
Material	*Inches*	*Centimeters*
Riveted steel	0.04–0.4	0.09–0.9
Concrete	0.01–0.1	0.02–0.2
Wood stove	0.007–0.04	0.02–0.09
Cast iron	0.0102	0.026
Galvanized iron	0.006	0.015
Asphalted cast iron	0.0048	0.012
Commercial steel or wrought iron	0.0018	0.046
PVC	0.000084	0.00021
Drawn tubing	0.00006	0.00015

Source: Analysis of Flow in Pipe Networks by Roland Jeppson. Used with permission of Ann Arbor Science Publishers.

A number of different expressions have been proposed to describe the relationship between f, Re, and e/D. Several of these relationships are presented in Table 1-2 and are summarized in graphical form as the Moody diagram (given as Fig. 1-7). The majority of flow situations encountered in the design of wastewater treatment plants are turbulent in nature. Jeppson (1977) indicates that Eq. (1-25) may be used to compute f for all turbulent flows. Equation (1-25) cannot be solved explicitly for f;

TABLE 1-2 *Summary of friction factor equations for Darcy–Weisbach equation, after Jeppson (1977).*

Type of flow	*Equation giving* f	*Range of application*	*Equation*
Laminar	$f = 64/\text{Re}$	Re < 2100	(1-22)
Hydraulically smooth or turbulent smooth	$f = 0.316/\text{Re}^{0.25}$	4000 < Re < 10^5	(1-23)
	$\dfrac{1}{\sqrt{f}} = 2 \log_{10}(\text{Re}\sqrt{f}) - 0.8$	Re > 4000	(1-24)
Transition between hydraulically smooth and wholly rough	$\dfrac{1}{\sqrt{f}} = 2\log_{10}\left(\dfrac{e/D}{3.7} + \dfrac{2.52}{\text{Re}\sqrt{f}}\right)$ $= 1.14 - 2\log_{10}\left(\dfrac{e}{D} + \dfrac{9.35}{\text{Re}\sqrt{f}}\right)$	Re > 4000	(1-25)
Hydraulically rough or turbulent rough	$\dfrac{1}{\sqrt{f}} = 1.14 - 2\log_{10}(e/D)$ $= 1.14 + 2\log_{10}(D/e)$	Re > 4000	(1-26)

Source: Analysis of Flow in Pipe Networks by Roland Jeppson. Used with permission of Ann Arbor Science Publishers.

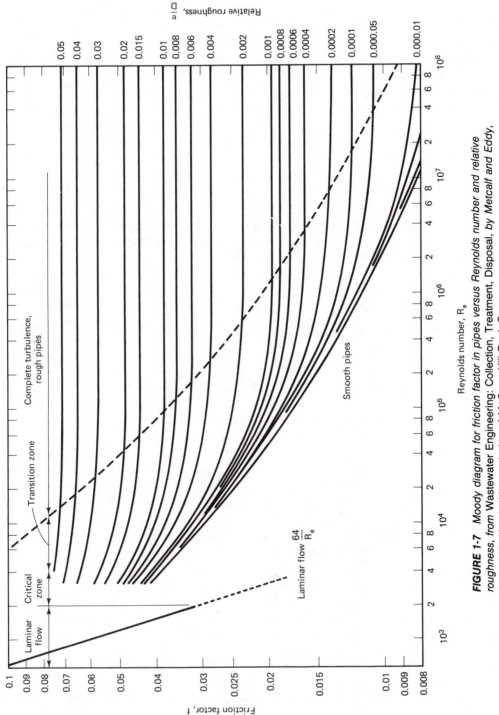

FIGURE 1-7 *Moody diagram for friction factor in pipes versus Reynolds number and relative roughness, from* Wastewater Engineering: Collection, Treatment, Disposal, *by Metcalf and Eddy, Inc., 1972. Used with permission of McGraw-Hill Book Company.*

however, an iterative procedure can be used to evaluate f for a particular flow situation. The usual procedure is to use Eq. (1-26) to obtain an initial estimate of f and then to adjust this value until the correct value of f is obtained from Eq. (1-25).

Wood (1966) has proposed an explicit relationship that may be used to approximate f for all turbulent flows. The appropriate expression is

$$f = a + b/(\text{Re})^c \tag{1-27}$$

where $a = 0.094(e/D)^{0.225} + 0.53(e/D)$
$b = 88(e/D)^{0.44}$
$c = 1.62(e/D)^{0.134}$

EXAMPLE PROBLEM 1-4:   The discharge through a straight section of 6-in. new cast-iron pipe is 0.5 cfs. What is the energy loss across 1000 ft of this pipe if the water temperature is 50°F?

Solution:

1. Determine e/D for a new 6-in. cast-iron pipe using Table 1-1 to obtain an estimate for e.

$$\frac{e}{D} = \frac{0.0102}{6} = \mathbf{0.0017}$$

2. Calculate the average flow velocity

$$Q = AV$$

or $$V = \frac{Q}{A} = \frac{0.5}{[\pi(6/12)^2/4]}$$

$$V = \mathbf{2.55 \ fps}$$

3. Compute the Reynolds number for the given flow conditions. Viscosity values for water are given in Table 1-3.

$$\text{For } 50°F: \ \nu = 1.41 \times 10^{-5} \ \text{ft}^2/\text{s}$$

$$\text{Re} = \frac{(2.55)(6/12)}{1.41 \times 10^{-5}}$$

$$= \mathbf{9 \times 10^4}$$

4. Estimate the friction factor from the Moody diagram given in Fig. 1-7.

$$f = \mathbf{0.024}$$

TABLE 1-3 *Properties of water.*

Temperature (°F)	Specific weight, γ (lb/ft³)	Mass density, ρ (lb-s²/ft⁴)	Dynamic viscosity, $\mu \times 10^5$ (lb-s/ft²)	Kinematic viscosity, $\nu \times 10^5$ (ft²/s)	Vapor pressure head P_V/γ (ft)
32	62.42	1.940	3.746	1.931	0.20
40	62.43	1.938	3.229	1.664	0.28
50	62.41	1.936	2.735	1.410	0.41
60	62.37	1.934	2.359	1.217	0.59
70	62.30	1.931	2.050	1.059	0.84
80	62.22	1.927	1.799	0.930	1.17
90	62.11	1.923	1.595	0.826	1.61
100	62.00	1.918	1.424	0.739	2.19
110	61.86	1.913	1.284	0.667	2.95
120	61.71	1.908	1.168	0.609	3.91
130	61.55	1.902	1.069	0.558	5.13
140	61.38	1.896	0.981	0.514	6.67
150	61.20	1.890	0.905	0.476	8.58
160	61.00	1.896	0.838	0.442	10.95
170	60.80	1.890	0.780	0.413	13.83
180	60.58	1.883	0.726	0.385	17.33
190	60.36	1.876	0.678	0.362	21.55
200	60.12	1.868	0.637	0.341	26.59
212	59.83	1.860	0.593	0.319	33.90

5. Calculate the energy loss from Eq. (1-20).

$$h_L = \frac{fLV^2}{2gD}$$

$$= \frac{(0.024)(1000)(2.55)^2}{(2)(32.2)(6/12)}$$

$$= \mathbf{4.8 \ ft}$$

Although the Darcy–Weisbach equation has the most rational base for pipe flow, two empirical equations are often employed for determining friction losses in pipe flow systems. These equations are the Manning equation and the Hazen–Williams equation. The Manning equation has the following form:

$$V = \frac{1.486}{n}R^{2/3}S^{1/2} \tag{1-28}$$

where V = average velocity of flow, ft/s

n = coefficient of roughness (see Table 1-4)

R = hydraulic radius (cross-sectional area of flow divided by wetted perimeter), ft

TABLE 1-4 *Values of n to be used with the Manning equation, after King and Brater (1976).*

Kind of pipe	Variation		Use in design	
	From	*To*	*From*	*To*
Clean uncoated cast iron	0.011	0.015	0.013	0.015
Clean coated cast iron	0.010	0.014	0.012	0.014
Tuberculated cast iron	0.015	0.035	—	—
Riveted steel	0.013	0.017	0.015	0.017
Lock bar and welded	0.010	0.013	0.012	0.013
Galvanized iron	0.012	0.017	0.015	0.017
Brass and glass	0.009	0.013	—	—
Concrete	0.010	0.017	—	—
Concrete with rough joints	—	—	0.016	0.017
Concrete, "dry mix," rough forms	—	—	0.015	0.016
Concrete, "wet mix," steel forms	—	—	0.012	0.014
Concrete, very smooth	—	—	0.011	0.012
Vitrified clay	0.010	0.017	0.013	0.015
Common clay drainage tile	0.011	0.017	0.012	0.014
Corrugated metal ($2\frac{2}{3}'' \times \frac{1}{2}''$)	0.023	0.026	—	—
Corrugated metal ($3'' \times 1''$ and $6'' \times 1''$)	0.026	0.029	—	—

Source: Handbook of Hydraulics, sixth ed. by Brater and King. Used with permission of McGraw-Hill Book Co.

S = slope of the energy line, i.e., the loss of energy per unit length of pipe, ft/ft.

This equation can be quickly solved through the use of the nomograph shown in Fig. 1-8.

The Hazen–Williams equation is probably used more often than any other equation for computing energy losses in pipe flow systems. The Hazen–Williams equation is

$$V = 1.318 \, CR^{0.63} S^{0.54} \qquad (1\text{-}29)$$

where V = average velocity in the pipe, ft/s

C = coefficient of roughness (see Table 1-5)

R = hydraulic radius, ft

S = slope of the energy line, ft/ft

A nomograph that can be used to facilitate the solution of Eq. (1-29) is given in Fig. 1-9. This equation is widely used in design and is most applicable for pipes 2 in. in diameter or larger and for velocities less than 10 fps. The major limitation of this equation is that temperature and viscosity variations are ignored.

Jain et al. (1978) indicate that the use of the common Hazen–Williams equation for computing energy loss in pipe flow systems may result in an appreciable error because the coefficient C is assumed to be independent of flow velocity, pipe diameter, and viscosity. These investigators propose a procedure for computing frictional losses based on the modified Hazen–Williams nomograph shown in Fig. 1-10 and the

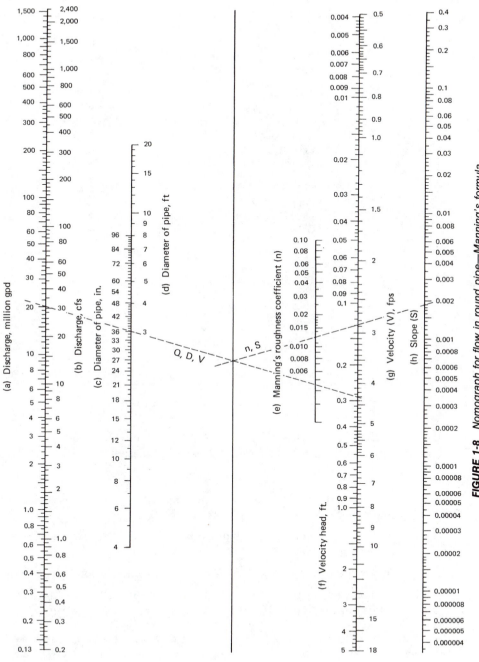

FIGURE 1-8 Nomograph for flow in round pipe—Manning's formula.

(a) Discharge, million gpd

(b) Discharge, cfs

(c) Diameter of pipe, in.

(d) Diameter of pipe, ft

(e) Manning's roughness coefficient (n)

(f) Velocity head, ft.

(g) Velocity (V), fps

(h) Slope (S)

Q, D, V

n, S

TABLE 1-5 *Values of C for the Hazen–Williams formula, after Simon (1976).*

Type of pipe	C
Asbestos cement	140
Brass	130–140
Brick sewer	100
Cast iron	
New, unlined	130
Old, unlined	40–120
Cement lined	130–150
Bitumastic enamel lined	140–150
Tar coated	115–135
Concrete or concrete lined	
Steel forms	140
Wooden forms	120
Centrifugally spun	135
Copper	130–140
Fire hose (rubber lined)	135
Galvanized iron	120
Glass	140
Lead	130–140
Plastic	140–150
Steel	
Coal-tar enamel lined	145–150
New unlined	140–150
Riveted	110
Tin	130
Vitrified clay	100–140

Source: Practical Hydraulics by Andrew L. Simon. Used with permission of John Wiley & Sons, Inc.

modified Hazen–Williams coefficients, C_R, presented in Table 1-6. The procedure for applying Fig. 1-10 and Table 1-6 to energy loss calculations is outlined in the following steps:

1. With the total flow in gallons per minute (gpm) and pipe diameter in inches, determine the flow velocity from Fig. 1-10.

2. With the flow velocity determined in Step 1, determine the appropriate C_R value from Table 1-6.

3. Divide the total flow used in Step 1 by C_R to obtain an adjusted flow.

4. Using the adjusted flow and actual pipe diameter, determine the head loss per 1000 ft of pipe length from Fig. 1-10.

EXAMPLE PROBLEM 1-5: Rework Example Problem 1-4 by applying the Manning equation, the Hazen–Williams equation, and the modified Hazen–Williams procedure. Compare the results with that given by the Darcy–Weisbach equation.

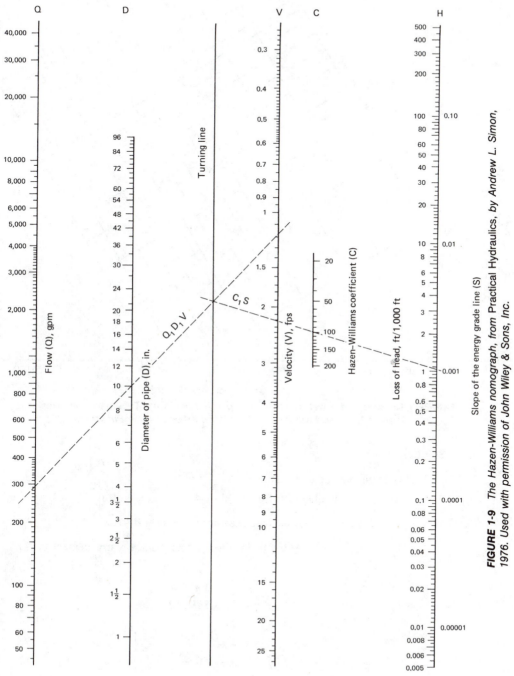

FIGURE 1-9 *The Hazen-Williams nomograph, from Practical Hydraulics, by Andrew L. Simon, 1976. Used with permission of John Wiley & Sons, Inc.*

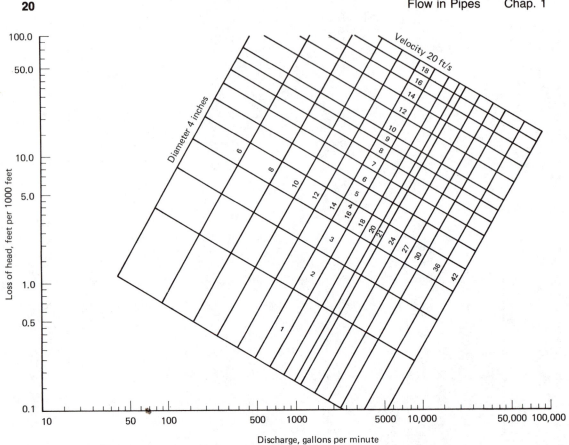

FIGURE 1-10 *Modified Hazen-Williams nomograph, from* Water Treatment Plant Design, *by Robert L. Sanks, 1978. Used with permission of Ann Arbor Science Publishers.*

Solution:

1. Manning equation calculations:

    (a) Obtain a value of n from Table 1-4.

        For uncoated cast-iron pipe $n = \mathbf{0.013.}$

    (b) Determine the slope of the energy line from a rearrangement of Eq. (1-28).

    $$S = \left[\frac{Vn}{1.486R^{2/3}} \right]^2$$

    $$= \left[\frac{(2.55)(0.013)}{1.486\left(\dfrac{6/12}{4}\right)^{2/3}} \right]^2$$

    $$S = \mathbf{0.0080 \ ft/ft}$$

TABLE 1-6   *Modified Hazen–Williams coefficient, C_R, after Amirtharajah (1978).*

Nominal diameter (in.)	Velocity, ft/sec						
	1.0	3.0	3.9	5.9	7.9	9.8	19.7
1. Cast-iron (new-coated) pipes							
4	0.9524	0.9089	0.8922	0.8659	0.8459	0.8299	0.7786
8	0.9760	0.9291	0.9117	0.8470	0.8641	0.8477	0.7953
12	0.9865	0.9379	0.9202	0.8927	0.8720	0.8554	0.8025
16	0.9927	0.9428	0.9249	0.8972	0.8763	0.8596	0.8064
20	0.9967	0.9459	0.9279	0.9000	0.8790	0.8623	0.8089
24	0.9950	0.9480	0.9299	0.9019	0.8809	0.8641	0.8106
28	1.0000	0.9495	0.9313	0.9032	0.8821	0.8653	0.8117
32	$C_R = 1.0$	0.9514	0.9323	0.9042	0.8837	0.8662	0.8125
36	$C_R = 1.0$	0.9524	0.9330	0.9049	0.8842	0.8668	0.8131
2. Cast-iron (old) pipes							
4	0.6552	0.6886	0.7170	0.5485	0.5325	0.5203	0.4841
8	0.7025	0.6311	0.6130	0.5881	0.5709	0.5578	0.5190
12	0.7261	0.6532	0.6335	0.6078	0.5900	0.5765	0.5364
16	0.6657	0.6466	0.6203	0.6203	0.6022	0.5884	0.5475
20	0.7518	0.6753	0.6559	0.6293	0.6293	0.5969	0.5554
24	0.7599	0.6826	0.6630	0.6361	0.6175	0.6034	0.5614
28	0.7664	0.6884	0.6687	0.6415	0.6227	0.6085	0.5661
32	0.7718	0.6932	0.6733	0.6459	0.6270	0.6127	0.5701
36	0.7762	0.6972	0.6772	0.6497	0.6307	0.6163	0.5734
3. Concrete (new) pipes							
4	1.0	1.0	0.9960	0.9821	0.9694	0.9581	0.9156
8	1.0	1.0	1.0	0.9909	0.9775	0.9657	0.9223
12	1.0	1.0	1.0	0.9935	0.9798	0.9677	0.9239
16	1.0	1.0	1.0	0.9943	0.9803	0.9682	0.9241
20	1.0	1.0	1.0	0.9944	0.9802	0.9681	0.9237
24	1.0	1.0	1.0	0.9936	0.9798	0.9675	0.9231
28	1.0	1.0	1.0	0.9930	0.9795	0.9668	0.9223
32	1.0	1.0	1.0	0.9923	0.9785	0.9661	0.9215
36	1.0	1.0	1.0	0.9916	0.9778	0.9653	0.9207
4. Steel (new) pipes							
4	0.9942	0.9766	0.9658	0.9466	0.9304	0.9166	0.8687
8	1.0	0.9905	0.9789	0.9587	0.9420	0.9279	0.8790
12	1.0	0.9973	0.9838	0.9631	0.9461	0.9318	0.8828
16	1.0	0.9985	0.9861	0.9651	0.9480	0.9336	0.8842
20	1.0	1.0	0.9873	0.9661	0.9488	0.9343	0.8848
24	1.0	1.0	0.9879	0.9665	0.9491	0.9346	0.8850
28	1.0	1.0	0.9881	0.9666	0.9492	0.9346	0.8849
32	1.0	1.0	0.9881	0.9662	0.9490	0.9344	0.8847
36	1.0	1.0	0.9880	0.9662	0.9487	0.9341	0.8844

Source: Water Treatment Plant Design by Robert L. Sanks, 1978. Used with permission of Ann Arbor Science Publishers.

(c) Calculate the total head loss for the 1000 ft length of pipe.

$$h_L = 0.0080 \text{ ft/ft} \times 1000 \text{ ft}$$
$$= \textbf{8.0 ft}$$

2. Hazen–Williams equation calculations:

(a) Obtain a value of C from Table 1-5 for uncoated cast-iron pipe $C = 130$.

(b) Determine the slope of the energy line from a rearrangement of Eq. (1-29).

$$S = \left[\frac{V}{1.318CR^{0.63}}\right]^{1.85}$$

$$= \left[\frac{2.55}{(1.318)(130)\left(\frac{6/12}{4}\right)^{0.63}}\right]^{1.85}$$

$$S = \textbf{0.0047 ft/ft}$$

(c) Calculate the total head loss for the 1,000 ft length of pipe.

$$h_L = 0.0047 \text{ ft/ft} \times 1000 \text{ ft}$$
$$= \textbf{4.7 ft}$$

3. Modified Hazen–Williams procedure:

(a) Convert flow from cfs to gpm

$$0.5 \frac{\text{ft}^3}{\text{s}} \times 7.48 \frac{\text{gal}}{\text{ft}^3} \times 60 \frac{\text{s}}{\text{min}} = \textbf{224.4 gpm}$$

(b) Estimate C_R from Table 1-6 (use new-coated cast-iron).

Interpolation gives a C_R of **0.9292**

(c) Calculate the adjusted flow by dividing Q in gpm by C_R.

$$Q_{adj} = \frac{224.4}{0.9292}$$

$$= \textbf{241.5 gpm}$$

(d) Using Q_{adj}, estimate the head loss in ft/1000 ft from Fig. 1-10.

$$h = \textbf{5.5 ft/1000 ft}$$

(e) Compute the total head loss for the 1000 ft length of pipe.

$$h_L = 5.5 \times \frac{\text{length of pipe}}{1000}$$

$$= 5.5 \times \frac{1000}{1000}$$

$$h_L = \mathbf{5.5 \ ft}$$

Comparing the computed head losses with that given by the Darcy–Weisbach equation shows that for this particular problem, the Hazen–Williams equation and the Darcy–Weisbach equation give comparable results. However, it should be understood that the Darcy–Weisbach equation always gives the most reliable answer.

1.3 MINOR LOSSES

If the velocity of a flowing fluid is changed either in direction or in magnitude, eddies are usually created, which results in a loss of energy in excess of that which would be realized if the energy loss over a specified length of pipe were due only to friction (e.g., expansions result in changes of magnitude, whereas flow through an elbow will result in a directional change of velocity). Energy losses due to local disturbances of the flow, such as those that occur at bends or valves, are called *minor losses*. The term minor losses is somewhat of a misnomer, since these types of losses can be quite significant and, in some cases, more important than frictional losses for relatively short flow reaches. Since most runs of pipe and open channels in wastewater treatment plants are fairly short, minor losses are important and must be considered when computing energy losses.

Minor losses are commonly expressed in terms of velocity head using the relationship

$$h_L = K\frac{V^2}{2g} = \left(\frac{8K}{\pi^2 g D^4}\right)Q^2 \qquad \textbf{(1-30)}$$

where K is referred to as the head loss coefficient and V is the average velocity of flow in the pipe *before* the flow passes through the appurtenance. Head loss coefficients for various appurtenances are given in Table 1-7.

EXAMPLE PROBLEM 1-6: Water at 70°F is to be delivered from tank A to tank B at the rate of 2 cfs by gravity flow. The pipeline conveying the fluid is to have an abrupt entrance, two 90° (short-radius) elbows, a gate valve, and a 1:2 orifice. The pipe will be 100 ft long and is to be made of new cast iron. Determine the diameter pipe needed to meet the discharge requirement.

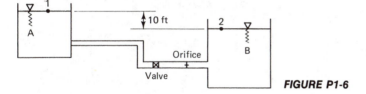

FIGURE P1-6

TABLE 1-7 *Head losses for various appurtenances [for the equation* $h_L = KV^2/2g$ *unless otherwise indicated, after Amirtharajah (1978)*].

Appurtenance (Alphabetically)	Head loss as multiple of $(V^2/2g)$	Appurtenance (Alphabetically)	Head loss as multiple of $(V^2/2g)$
1. Butterfly valves		Pipe flush with tank	0.5
Fully open	0.3	Pipe projecting into tank	
Angle closed, $\Theta = 10°$	0.46	(Borda entrance)	0.83–1.0
$\Theta = 20°$	1.38	Slightly rounded	0.23
$\Theta = 30°$	3.6	Strainer and foot valve	2.50
$\Theta = 40°$	10	**9. Gate valves**	
$\Theta = 50°$	31	Open	0.19
$\Theta = 60°$	94	$\frac{1}{4}$ closed	1.15
2. Check (reflux) valves		$\frac{1}{2}$ closed	5.6
Ball type (fully open)	2.5–3.5	$\frac{3}{4}$ closed	24.0
Horizontal lift type	8–12	also see Sluice gates	
Swing check	0.6–2.3	**10. Increasers**	
Swing check (fully open)	2.5	$0.25\,(V_1^2/2g - V_2^2/2g)$	
3. Contraction—sudden		where V_1 = velocity at small end	
4:1 (in terms of velocities		**11. Miter bends**	
of small end)	0.42	Deflection angle, Θ	
2:1	0.33	5°	0.016–0.024
4:3	0.19	10°	0.034–0.044
also see Reducers		15°	0.042–0.062
4. Diaphragm valve		22.5°	0.066–0.154
Fully open	2.3	30°	0.130–0.165
$\frac{3}{4}$ open	2.6	45°	0.236–0.320
$\frac{1}{2}$ open	4.3	60°	0.471–0.684
$\frac{1}{4}$ open	21.0	90°	1.129–1.265
5. Elbow—90°		**12. Obstructions in pipes** (in terms of pipe	
Flanged–regular	0.21–0.30	velocities) pipe to obstruction area ratio	
Flanged–long radius	0.18–0.20	1.1	0.21
Intersection of two cylinders		1.4	1.15
(welded pipe–		1.6	2.40
not rounded)	1.25–1.8	2.0	5.55
Screwed–short radius	0.9	3.0	15.0
Screwed–medium radius	0.75	4.0	27.3
Screwed–long radius	0.60	5.0	42.0
6. Elbow—45°		6.0	57.0
Flanged–regular	0.20–0.30	7.0	72.5
Flanged–long radius	0.18–0.20	10.0	121.0
Screwed–regular	0.30–0.42	**13. Orifice meters** (in terms of velocities of	
7. Enlargement—sudden		pipe) orifice to pipe diameter ratio	
1:4 (in terms of velocities		0.25 (1:4)	4.8
of small end)	0.92	0.33 (1:3)	2.5
1:2	0.56	0.50 (1:2)	1.0
3:4	0.19	0.67 (2:3)	0.4
also see Increasers		0.75 (3:4)	0.24
8. Entrance losses			
Bell mouthed	0.04		

TABLE 1-7   *(cont.)*

Appurtenance (Alphabetically)	*Head loss as multiple of* $(V^2/2g)$	*Appurtenance (Alphabetically)*	*Head loss as multiple of* $(V^2/2g)$
14. *Outlet losses*		19. *Tees*	
Bell-mouthed outlet	$0.1\left(\dfrac{V_1^2}{2g} - \dfrac{V_2^2}{2g}\right)$	Standard–bifurcating	1.5–1.8
		Standard–90° turn	1.80
Sharp-cornered outlet	$\left(\dfrac{V_1^2}{2g} - \dfrac{V_2^2}{2g}\right)$	Standard–run of tee	0.60
		Reducing–run of tee	
Pipe into still water or air		2:1 (based on velocities	0.90
(free discharge)	1.0	4:1 of smaller end)	0.75
15. *Plug globe or stop valve*		20. *Venturi meters*	
Fully open	4.0	The head loss occurs mostly in and	
$\frac{3}{4}$ open	4.6	downstream of throat, but losses shown	
$\frac{1}{2}$ open	6.4	are given *in terms of velocities at inlet*	
$\frac{1}{4}$ open	780.0	*ends to assist in design.*	
16. *Reducers*		Long tube type—throat-to-inlet diameter	
Ordinary (in terms of velocities		ratio	
of small end)	0.25	0.33 (1:3)	1.0–1.2
Bell mouthed	0.10	0.50 (1:2)	0.44–0.52
Standard	0.04	0.67 (2:3)	0.25–0.30
Bushing or coupling	0.05–2.0	0.75 (3:4)	0.20–0.23
17. *Return bend (2 Nos. 90°)*		Short tube type—throat-to-inlet diameter	
Flanged–regular	0.38	ratio	
Flanged–long radius	0.25	0.33 (1:3)	2.43
Screwed	2.2	0.50 (1:2)	0.72
18. *Sluice gates*		0.67 (2:3)	0.32
Contraction in conduit	0.5	0.75 (3:4)	0.24
Same as conduit width without			
top submergence	0.2		
Submerged port in 12-in. wall	0.8		

Source: Water Treatment Plant Design by Robert L. Sanks, 1978. Used with permission of Ann Arbor Science Publishers.

Solution:

1. Write the Bernoulli equation between points 1 and 2 located on the water surface of tanks A and B, respectively.

$$\frac{p_1}{\gamma} + Z_A + \frac{V_1^2}{2g} = \frac{p_2}{\gamma} + Z_B + \frac{V_2^2}{2g} + (\Sigma K)\frac{V_p^2}{2g} + \frac{fL}{D}\frac{V_p^2}{2g}$$

Where V_p represents the average velocity of flow in the pipe and ΣK represents the sum of the minor head loss coefficients.

2. From the Bernoulli equation given in Step 1, develop an equation that gives the relationship between f and D.

    The pressure head and velocity head terms at the water surface in the tanks are zero; therefore, Bernoulli's equation reduces to

    $$(Z_A - Z_B) = (\Sigma K)\frac{V_p^2}{2g} + \frac{fL}{D}\frac{V_p^2}{2g}$$

    or

    $$(Z_A - Z_B) = \left[\frac{8(\Sigma K)}{\pi^2 gD^4} + \frac{8fL}{\pi^2 gD^5}\right]Q^2$$

    $$10 = \left[\frac{8(0.5 + 0.9 + 0.9 + 0.19 + 0.4 + 1.0)}{\pi^2(32.2)D^4} + \frac{8f(100)}{\pi^2(32.2)D^5}\right](2)^2$$

    Rearranging gives

    $$2.5D^5 = 2.517f + 0.0979D$$

    or

    $$D = (1.007f + 0.0392D)^{0.2}$$

3. Solve the relationship between D and f developed in Step 2 by the following iterative procedure:

    (a) Assume a value of f.

    $$f = 0.03$$

    (b) Calculate D from the relationship given in Step 2 by trial and error.

    $$D = (1.007 \, (0.03) + 0.0392D)^{0.2}$$
    $$D = \mathbf{0.55 \ ft}$$

    (c) Compute the Reynolds number.

    $$Re = \frac{4Q}{\pi}\frac{1}{D}$$

    $$= \frac{4(2)}{(1.059 \times 10^{-5})}\left(\frac{1}{0.55}\right)$$

    $$= \mathbf{1.75 \times 10^6}$$

    (d) Determine the e/D value.

    $$e/D = \frac{.0102}{(0.55)(12)} = \mathbf{0.00155}$$

    (e) Read a new f value from the Moody diagram (Fig. 1-7) using Re and e/D from Steps (c) and (d).

$$f = \mathbf{0.022}$$

(f) If the new f value equals the last f value when rounded to two significant
digits, the calculated D is correct. If not, repeat Steps (b) through (f)
using the newest f value obtained in Step (c).

$$0.022 \neq 0.03$$

Therefore, repeat Steps (b) through (f) using an f value of 0.022. After
one more iteration, it is found that $D = \mathbf{0.53\ ft}$ or $\mathbf{6.36\ in.}$

Note: This is the minimum pipe size that can be used. The design
engineer would normally select a larger standard pipe size and
adjust the valve to give the desired flow.

Pipes in Series

The flow through a pipeline consisting of several different pipes connected in series
can be analyzed by assuming hydraulically rough flow and developing a head loss
equation for the entire pipeline, including appurtenances. This can be done by following the procedure:

1. Determine the f value as a function of only e/D for each pipe using Fig. 1-7
   or Eq. (1-26).

2. Determine the sum of the minor head loss coefficients for each pipe using the
   values in Table 1-7.

3. Determine the discharge factor for each pipe according to the equation

$$h_i = G_i Q_i^2 = \left[\frac{8 f_i L_i}{\pi^2 g D_i^5} + \frac{8(\Sigma K)_i}{\pi^2 g D_i^4} \right] Q_i^2$$

where
$$h_i = \text{head loss through } i\text{th pipe,}$$
$$G_i = \text{discharge factor for } i\text{ th pipe,}$$
$$Q_i = \text{discharge through } i\text{th pipe,}$$
$$(\Sigma K)_i = \text{sum of minor head loss coefficients for } i\text{th pipe,}$$
$$f_i, L_i, \text{ and } D_i = \text{friction factor, length, and diameter, respectively,}$$
$$\text{of } i\text{th pipe.}$$

4. Determine the discharge factor for the pipe series.
   Since the discharge is the same through each pipe in the series and the total
   head loss through the system equals the sum of the head losses for each pipe,
   the discharge factor is determined according to the equation

$$h_s = \left[\sum_{i=1}^{n} G_i \right] Q^2 = G_s Q^2$$

where h_s = total head loss through system

             n = number of pipes in series

             G_s = series discharge coefficient

             Q = total discharge

5. Determine discharge or head loss for the system from the relationship in Step 4.

    After determining Q or h_s in Step 5, one may want to evaluate the validity of the hydraulically rough flow assumption by checking f values from the Moody diagram using both e/D and the Reynolds number. If they differ significantly from the initial f values, the analysis should be repeated with the improved friction factors.

Pipes in Parallel

When two or more pipes are in parallel, it is also useful to develop a head loss equation for the pipe system. This is accomplished as follows:

1 through 3.   Same as for pipes in series analysis.

4. Detemine the discharge factor for the parallel pipes system. For parallel pipes, the head loss, h, is the same through each pipe but the total discharge, Q, is the sum of the discharges through each of the pipes. Therefore, the head loss equation is developed as follows

$$Q = Q_1 + Q_2 + \ldots + Q_n$$

$$Q = \sqrt{\frac{h}{G_1}} + \sqrt{\frac{h}{G_2}} + \ldots + \sqrt{\frac{h}{G_n}}$$

$$Q = \left[\sum_{i=1}^{n} \sqrt{1/G_i} \right] \sqrt{h}$$

or, rearranging,

$$h = \frac{1}{\left[\sum_{i=1}^{n} \sqrt{1/G_i} \right]^2} Q^2 = G_p Q^2$$

where h = head loss through system,

             n = number of pipes in parallel, and

$$G_p = \frac{1}{\left[\sum_{i=1}^{n} \sqrt{1/G_i}\right]^2} = \text{parallel discharge factor}$$

5. Same as for pipes in series analysis.

EXAMPLE PROBLEM 1-7:   A multiple pipe system connects tanks A and B as shown below. The diameter, length, equivalent roughness, and sum of minor head loss coefficients are given in Columns 2–5 of the computation table for each of the eight pipes in the system. If the water surface elevations in tanks A and B are 55 and 10 ft, respectively, determine the flow rate from tank A to tank B and the discharge in pipe 4. Assume hydraulically rough flow in all pipes.

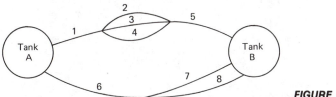

FIGURE P1-7

Computation Table for Example Problem 1-7.

Pipe no.	Diameter D (ft)	Length L (ft)	Equivalent roughness e (ft)	Sum of minor head loss coefficients (ΣK)	Relative roughness (e/D)	Friction factor (f)	Discharge coefficient (G)
(1)	(2)	(3)	(4)	(5)	(6)	(7)	(8)
1	0.75	200	0.001	1.5	0.00133	0.0211	0.567
2	0.50	130	0.001	2.5	0.00200	0.0234	3.457
3	0.50	75	0.002	1.4	0.00400	0.0284	2.280
4	0.75	145	0.002	2.5	0.00267	0.0253	0.588
5	1.00	150	0.001	1.4	0.00100	0.0196	0.109
6	0.50	230	0.0015	1.6	0.00300	0.0261	5.480
7	0.50	250	0.0025	2.1	0.00500	0.0303	6.948
8	0.75	260	0.0015	2.1	0.00200	0.0234	0.812

Solution:

1. Determine the f values using Eq. (1-26). The e/D values and the corresponding f values are given in Columns 6 and 7 of the computation table.

2. Determine the sum of the minor head loss coefficients for each pipe. These values are given in column 5 of the computation table.

3. Determine the discharge factor for each pipe using the equation

$$G_i = \frac{8f_i L_i}{\pi^2 g D_i^5} + \frac{8(\Sigma K)_i}{\pi^2 g D_i^4} = \frac{1}{39.73}\left[\frac{f_i L_i}{D_i^5} + \frac{(\Sigma K)_i}{D_i^4}\right]$$

The G values are calculated as given in Column 8 of the computation table.

4. Determine the parallel discharge factors for pipes 2, 3, and 4 and for pipes 7 and 8.

   (a) Pipes 2, 3, and 4

$$G_{P_{2-4}} = \frac{1}{\left[\sum\limits_{i=2}^{4} \sqrt{1/G_i}\right]^2} = \frac{1}{[\sqrt{1/3.457} + \sqrt{1/2.280} + \sqrt{1/0.580}]^2}$$

$$G_{P_{2-4}} = \mathbf{0.159}$$

   (b) Pipes 7 and 8

$$G_{P_{7,8}} = \frac{1}{\left[\sum\limits_{i=7}^{8} \sqrt{1/G_i}\right]^2} = \frac{1}{[\sqrt{1/6.948} + \sqrt{1/0.812}]^2}$$

$$G_{P_{7,8}} = \mathbf{0.451}$$

5. Determine the series discharge coefficients for pipes 1, 2, 3, 4, and 5 and for pipes 6, 7, and 8.

   (a) Pipes 1, 2, 3, 4, and 5

$$G_{S_{1-5}} = G_1 + G_{P_{2-4}} + G_5 = 0.567 + 0.159 + 0.109$$

$$G_{S_{1-5}} = \mathbf{0.835}$$

   (b) Pipes 6, 7, and 8

$$G_{S_{6-8}} = G_6 + G_{P_{7,8}} = 5.480 + .451$$

$$G_{S_{6-8}} = \mathbf{5.931}$$

6. Determine the parallel discharge coefficient for all eight pipes. (This is the system discharge coefficient.)

$$G_{\text{system}} = G_{P_{1-8}} = \frac{1}{[\sqrt{1/G_{1,2,3,4,5}} + \sqrt{1/G_{6,7,8}}]^2}$$

$$= \frac{1}{[\sqrt{1/.835} + \sqrt{1/5.931}]^2}$$

$$G_{\text{system}} = \mathbf{0.442}$$

7. Determine the discharge from tank A to tank B.

$$Q = \sqrt{h/G_{system}} = \sqrt{\frac{(Z_A - Z_B)}{G_{system}}}$$

$$Q = \sqrt{\frac{55 - 10}{0.442}} = \textbf{10.09 cfs}$$

8. Determine the discharge through pipe 4.

    (a) Determine the flow through the piping consisting of pipes 1, 2, 3, 4, and 5

    $$Q_{1-5} = \sqrt{h/G_{1-5}} = \sqrt{\frac{Z_A - Z_B}{G_{1-5}}} = \sqrt{\frac{55 - 10}{0.835}}$$

    $$Q_{1-5} = \textbf{7.34 cfs}$$

    (b) Determine the head loss in the parallel piping containing pipes 2, 3, and 4

    $$h_{2-4} = Z_A - h_1 - h_5 - Z_B$$

    $$= 55 - G_1 Q_{1-5}^2 - G_5 Q_{1-5}^2 - 10$$

    $$h_{2-4} = 55 - 0.567(7.34)^2 - 0.109(7.34)^2 - 10$$

    $$h_{2-4} = 8.58 \text{ ft}$$

    Since the head loss is the same through each of pipes 2, 3, and 4, the discharge through pipe 4 is

    $$Q_4 = \sqrt{\frac{h_4}{G_4}} = \sqrt{\frac{8.58}{0.588}} = \textbf{3.82 cfs}$$

1.4 SLUDGE FLOW

The hydraulic analysis of a pipeline carrying concentrated sludge is often a difficult matter because the material is a non-Newtonian fluid. Vesilind (1979) indicated that the two methods commonly used for estimating energy loss in a pipe carrying sludge are the Hazen–Williams equation [Eq. (1-29)] with a modified C value and a graphical method based on field data. When the Hazen–Williams equation is used, the coefficient, C, is decreased, because concentrated sludge is more difficult to pump than water. A correlation between C and solids' content is presented in Table 1-8. It should be noted that dilute suspensions such as activated sludge behave no differently from water. Only after the solids' concentration exceeds 1% should adjustments be made for differences in fluid behavior.

In some cases the use of the Hazen–Williams equation for computing head losses results in the use of pipe diameters that are unnecessarily large. Because of this, many engineers simply try to provide pipe diameters that will result in average flow veloc-

TABLE 1-8   *Hazen–Williams coefficients for various solids' concentrations of raw sludge, after Brisbin (1957).*

Total solids (%)	Apparent H–W coefficient, C, based on C = 100 for water
0	100
2	81
4	61
6	45
8.5	32
10	25

ities between 5 and 8 ft/sec. Vesilind (1979) has presented a figure (see Fig. 1-11) that gives the relationship between head loss and average flow velocity for various solids' concentrations. This figure was developed from experimental data and should be fairly representative of what may be expected under field conditions.

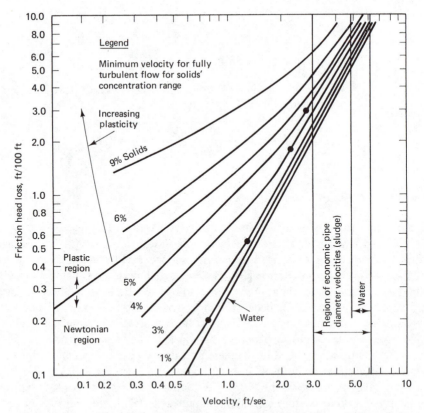

FIGURE 1-11   *Head loss versus velocity for various solids' concentrations, from* Treatment and Disposal of Wastewater Sludges, *second edition, by P. Aarne Vesiland, 1979. Used with permission of Ann Arbor Science Publishers.*

1.5 DIVIDING-FLOW MANIFOLDS

A common application of dividing-flow manifolds in environmental engineering is to distribute flow uniformly across the width of one or several treatment units (e.g., three parallel rectangular sedimentation basins, as illustrated in Fig. 1-12). The advantage of having a flow arrangement as shown in Fig. 1-12 was given by Camp (1961), who stated, ". . . where there are more than two rectangular tanks in parallel with common division walls and individual effluent weirs, perfect distribution requires that all effluent weirs be alike and at the same level and that each tank be fed by a separate conduit from a common junction point. Since the inlet conduits will differ in length, they must also differ in cross section in order to be identical hydraulically. It is a difficult procedure in design to make the inlet conduits identical hydraulically, and the use of a separate line to each tank is unduly expensive." Hence, Camp recommends the flow system presented in Fig. 1-12. However, perfect flow distribution for this arrangement requires that the head loss across each port be the same if all ports are

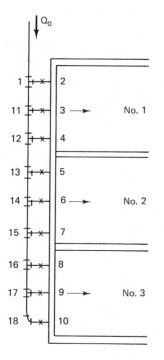
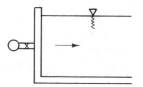

FIGURE 1-12 *Flow distribution between parallel rectangular sedimentation basins, after Camp (1961).*

the same diameter, i.e.,

$$(\Delta h)_{1,2} = (\Delta h)_{11,3} = \ldots = (\Delta h)_{18,10}$$

The design of this type of system is also quite difficult. To simplify things, Camp (1961) suggested that the design should allow a *small* flow variation to occur. Such an approach is normally acceptable, since *perfect* flow distribution is not essential for high treatment efficiency. The discussion presented by Camp (1961) will be summarized here.

Assume that the laterals have the same size and shape and that the relationship between discharge and head loss for each lateral is the same as that for an orifice, mouthpiece (a short piece of pipe whose length is about three times its diameter), or nozzle, i.e.,

$$q = C_d a \sqrt{2g\Delta h} \tag{1-31}$$

where

q = discharge through the lateral, cfs

C_d = discharge coefficient

a = cross-sectional area of lateral, ft^2

g = gravity constant, ft/s^2

Δh = the difference between the elevation of the energy line on each side of the lateral, ft
This is also the total head loss across the lateral.

Equation (1-31) may be solved for Δh to give

$$\Delta h = \left(\frac{1}{2gc_d^2 a^2}\right) q^2 \tag{1-32}$$

Since g and a are constants and c_d is assumed constant, the bracketed term on the right-hand side of Eq. (1-32) may be expressed as a constant K (since each lateral is assumed to have the same size and shape, K will be the same for each one).

$$\Delta h = Kq^2 \tag{1-33}$$

Thus, the head loss at point 2 between points 1 and 2 (see Fig. 1-12) is given by

$$(\Delta h)_{1-2} = K(q)^2_{1-2} \tag{1-34}$$

Flow control at the exit end of treatment units, such as sedimentation basins and aeration tanks, is normally achieved with weirs. The head on a weir sets the controlling depth–discharge relationship. Flow schematics for sharp-crested weirs and V-notch weirs are shown in Fig. 1-13. The relationship between discharge and head

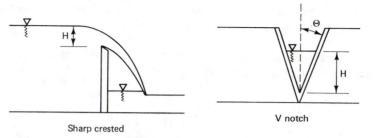

Sharp crested

V notch

FIGURE 1-13   *Weir flow.*

for sharp-crested weirs with free discharge is

$$H = \left[\frac{1}{C_W^{2/3} L^{2/3}}\right] Q^{2/3} \qquad (1\text{-}35)$$

where H = head on weir, ft

   C_W = weir coefficient

   L = length of the weir, ft

   Q = total discharge over the weir, cfs

Since C_W and L are constant for a particular situation, Eq. (1-35) may be expressed as

$$H = K_w Q^{2/3} \qquad (1\text{-}36)$$

where K_w represents the bracketed term on the right-hand side of Eq. (1-35).

The relationship between discharge and head for V-notch weirs with free discharge is

$$H = \left[\frac{1}{\left(C_W \tan \frac{\Theta}{2}\right)^{2/5}}\right] Q^{2/5} \qquad (1\text{-}37)$$

or

$$H = K_v Q^{2/5} \qquad (1\text{-}38)$$

where K_v represents the bracketed term on the right-hand side of Eq. (1-37).

If it is assumed that a sharp-crested weir is employed for flow control in each of the tanks shown in Fig. 1-12, the head on the weir in tank No. 1 is given by

$$H_1 = K_w (nq_{1-2})^{2/3} \qquad (1\text{-}39)$$

where n = number of laterals to tank No. 1

q_{1-2} = discharge through first lateral where it is assumed that the discharge through each lateral to tank No. 1 is the same, cfs.

If M represents some *arbitrarily selected* allowable flow variation between the *first lateral* and the *last lateral*, then from Fig. 1-12

$$M = \frac{q_{1-2}}{q_{18-10}} \tag{1-40}$$

It is assumed that the header pipe is horizontal and that the effluent weir in each sedimentation tank is set at the same level. One possible situation for head loss between the first lateral and the last lateral is illustrated in Fig. 1-14. In this figure the reference datum is at the top of the effluent weir. The graphical illustration presented in Fig. 1-14 suggests that

$$h_L = [(\Delta h)_{1-2} + H_1 + H_R] - [(\Delta h)_{18-10} + H_3 + H_R]$$

or

$$h_L = (\Delta h)_{1-2} - (\Delta h)_{18-10} + H_1 - H_3$$

or

$$h_L = (\Delta h)_{first} - (\Delta h)_{last} + H_{first} - H_{last} \tag{1-41}$$

where

h_L = the difference in head available to the first lateral and the last lateral, ft

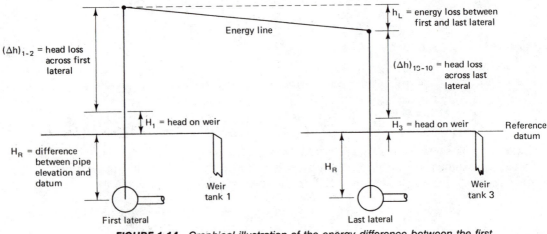

FIGURE 1-14 *Graphical illustration of the energy difference between the first and last lateral.*

It is important to note that one of two situations can exist in a dividing-flow manifold system:

1. The manifold may be quite small so that the head loss due to friction along the manifold is large. In this case the flow through the first lateral will probably be greater than the flow through the last lateral. The h_L in Eq. (1-41) will be positive, and M will have a value greater than unity.

2. The manifold may be quite large so that the head loss due to friction along the manifold is small. In this case velocity head recovery occurs along the manifold due to the decreasing flow rate. Thus, the presure will be greater at the downstream end of the manifold and, according to Eq. (1-31), this will result in a higher flow through the last lateral than through the first lateral. The h_L in Eq. (1-41) will be negative and M will have a value less than unity.

Making the appropriate substitutions, Camp (1961) showed that when sharp-crested weirs are used for effluent flow control, the head loss through the first lateral must be

$$(\Delta h)_{1-2} = \left[\frac{M^2}{M^2 - 1}\right]h_L - \left[M^{4/3}\left(\frac{M^{2/3} - 1}{M^2 - 1}\right)H_1\right] \qquad \textbf{(1-42)}$$

When V-notch weirs are used, the appropriate relationship is

$$(\Delta h)_{1-2} = \left[\frac{M^2}{M^2 - 1}\right]h_L - \left[M^{8/5}\left(\frac{M^{2/5} - 1}{M^2 - 1}\right)H_1\right] \qquad \textbf{(1-43)}$$

If the header pipe discharges into only one basin, the last terms in Eqs. (1-42) and (1-43) become zero, and each of these equations reduces to

$$(\Delta h)_{1-2} = \left[\frac{M^2}{M^2 - 1}\right]h_L \qquad \textbf{(1-44)}$$

It should be noted that if the design is made with a selected value of M at maximum flow, the variation in distribution will be less at smaller flow rates.

Fair (1951) notes that it is possible to approximate equal flow distribution by designing the system so that the head loss along the header pipe is small in comparison to the head loss across an individual lateral. Rich (1961) summarized the work of Fair (1951), and his discussion will be outlined in the following paragraphs.

Consider Fig. 1-15, which shows the head loss along a dividing-flow manifold when the flow along its length decreases uniformly. If the flow through the laterals is assumed to vary slightly, then it is possible to represent the relationship between the discharge through the first lateral and the discharge through the last lateral in the form

$$q_N = \psi q_1 \qquad \textbf{(1-45)}$$

where q_N = discharge through the last lateral, cfs

q_1 = discharge through the first lateral, cfs

ψ = ratio of the discharge through the last lateral to the discharge through the first lateral.

If the laterals are considered to be mouthpieces or orifices, Eq. (1-32) suggests that head loss through the first and last lateral may be formulated in terms of discharge as

$$(\Delta h)_1 = K(q_1)^2 \qquad\qquad\qquad \textbf{(1-46)}$$

$$(\Delta h)_N = K(q_N)^2 \qquad\qquad\qquad \textbf{(1-47)}$$

If q_N is expressed in terms of q_1 [see Eq. (1-45)], Eq. (1-47) may be written as

$$(\Delta h)_N = K(\psi q_1)^2 = \psi^2 K(q_1)^2 \qquad\qquad \textbf{(1-48)}$$

Substituting from Eq. (1-46) for $K(q_1)^2$ in Eq. (1-48) gives

$$(\Delta h)_N = \psi^2(\Delta h)_1 \qquad\qquad\qquad \textbf{(1-49)}$$

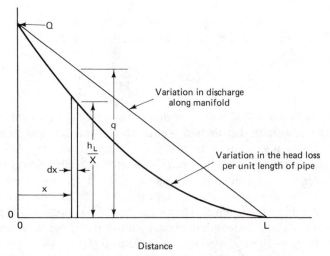

FIGURE 1-15   *Head loss when flow varies uniformly along pipe length, after Fair (1951).*

Since h_L is the head loss per unit length of pipe, it may be substituted for h_L/x in Eq. (1-55), resulting in the equation

$$(h_L)_x = \int_0^x K\left[Q_0\left(\frac{L - x}{L}\right)\right]^2 dx$$

or

$$(h_L)_x = \frac{KQ_0^2}{(L)^2} \int_0^x (L - x)^2 dx \tag{1-60}$$

Equation (1-60) may be integrated to give

$$(h_L)_x = \frac{KQ_0^2}{(L)^2}\left[L^2x - Lx^2 + \frac{1}{3}x^2 \right]$$

or

$$(h_L)_x = KQ_0^2\left[x - \frac{x^2}{L} + \frac{1}{3}\frac{x^3}{(L)^2} \right] \tag{1-61}$$

If no flow were lost in the manifold, i.e., if it were simply a straight pipe with no laterals, the head loss per unit of pipe length would be

$$\frac{h_T}{L} = KQ_0^2 \tag{1-62}$$

or

$$Q_0^2 = \frac{h_T}{LK} \tag{1-63}$$

where h_T = head loss when the total flow passes through the pipe, ft

Substituting for Q_0^2 in Eq. (1-61) from Eq. (1-63) gives

$$(h_L)_x = h_T\left[\frac{x}{L} - \left(\frac{x}{L}\right)^2 + \frac{1}{3}\left(\frac{x}{L}\right)^3 \right] \tag{1-64}$$

When x equals L, Eq. (1-64) reduces to

$$h_L = \frac{1}{3}h_T \tag{1-65}$$

This relationship implies that the head loss between the first and last lateral in the manifold is approximately $\frac{1}{3}$ of the head loss, which would be observed if the total flow passed through the manifold.

EXAMPLE PROBLEM 1-8: Determine the size of the mouthpiece required to distribute 10 million gpd of flow between two sedimentation tanks evenly.

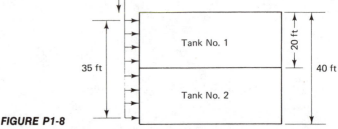

FIGURE P1-8

The tanks are 20 ft wide and flow control is achieved by sharp-crested weirs, which extend the width of each tank. A $q_N/q_1 = 1.5$ flow variation ratio between the first inlet and the last inlet is acceptable. A 12-in. pipe is to be used in the design.

Solution:

1. Compute the head on the weir in tank No. 1.

    (a) Determine ψ and M based on the given flow variation between the first and last inlet.

    $$\frac{q_N}{q_1} = \psi = \textbf{1.5}$$

    $$\frac{q_1}{q_N} = M = \textbf{0.67}$$

    (b) Calculate the head on the weir in tank No. 1.

    $$H_1 = \left[\frac{1}{C_W^{2/3}L^{2/3}}\right]Q^{2/3}$$

    For a sharp-crested weir where the length of the weir spans the width of the channel and where the approach velocity of flow can be neglected, $C_W \simeq 3.33$.

    $$H_1 = \left[\frac{1}{(3.33)^{2/3}(20)^{2/3}}\right]\left[4\left(\frac{10 \times 1.547}{8}\right)\right]^{2/3}$$

    $$= \textbf{0.24 ft}$$

    Note: Assume equal flow distribution for this calculation.

2. Determine the head loss through the manifold if it were passing the entire flow. Use the Hazen–Williams equation (Eq. (1-29)), and assume a C of 130.

$$S = \left[\frac{V}{1.318CR^{0.63}}\right]^{1.85}$$

$$= \left[\frac{4(10 \times 1.547)/\pi(1)^2}{1.318(130)\left(\frac{1}{4}\right)^{0.63}}\right]^{1.85}$$

$$= 0.092 \ \text{ft/ft}$$

Therefore,

$$h_T = 0.092 \times 35 \ \text{ft}$$

$$= \mathbf{3.2 \ ft}$$

3. From Eq. (1-65) estimate the actual head loss between the first and last lateral.

$$h_L = \frac{1}{3}h_T$$

$$= \frac{3.2}{3} = \mathbf{1.067 \ ft}$$

Note: Since M is less than 1.0, h_L should be given a negative sign.

4. Estimate the head loss through the first lateral by applying Eq. (1-42).

$$(\Delta h)_{1-2} = \left[\frac{M^2}{M^2 - 1}\right]h_L - \left[M^{4/3}\left(\frac{M^{2/3} - 1}{M^2 - 1}\right)H_1\right]$$

$$= \left[\frac{(0.67)^2}{(0.67)^2 - 1}\right](-1.067) - \left[(0.67)^{4/3}\left(\frac{(0.67)^{2/3} - 1}{(0.67)^2 - 1}\right)0.24\right]$$

$$= 0.87 - 0.06$$

$$= \mathbf{0.81 \ ft}$$

5. Determine the required lateral size by applying Eq. (1-31) and assuming $C_d = 0.82$.

$$q = C_d a \sqrt{2g \Delta h}$$

or

$$a = \frac{q}{C_d \sqrt{2g \Delta h}}$$

$$= \frac{(1.547 \times 10)/8}{(0.82)\sqrt{(2)(32.2)(0.81)}}$$

$$= 0.326 \ \text{ft}^2$$

Then,

$$D = \sqrt{\frac{4a}{\pi}}$$

$$= \sqrt{\frac{4 \times 0.326}{\pi}}$$

= 0.64 ft or 7.7 in.

A somewhat different approach to the hydraulic analysis of dividing-flow manifolds has been given by Hudson et al. (1979). These workers presented the schematic for a dividing-flow manifold shown in Fig. 1-16 and indicated that the total head loss from point 1 to point 2 was the sum of

1. The friction loss in the manifold,

2. The entrance loss to the lateral,

3. The friction loss in the lateral, and

4. The exit loss from the lateral.

Since the friction losses are normally insignificant when compared to the minor losses, the total head loss between points 1 and 2 may be approximated from the relationship

$$\Delta h = h_E + \frac{V_L^2}{2g} \qquad (1\text{-}66)$$

where Δh = head loss between points 1 and 2, ft

h_E = lateral entry loss, ft

$\dfrac{V_L^2}{2g}$ = lateral exit loss, ft

The entry loss term, h_E, may be expressed as

$$h_E = \frac{\alpha V_L^2}{2g} \qquad (1\text{-}67)$$

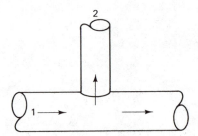

FIGURE 1-16 *Dividing-flow manifold.*

where α = entry loss coefficient, which is a function of the ratio of velocity in the manifold upstream from the lateral to the velocity in the lateral plus a constant, Θ.

According to Hudson et al. (1979), a linear relationship exists between α and the ratio of the average flow velocity in the manifold upstream from the lateral to the average flow velocity in the lateral. This relationship may be expressed mathematically as

$$\alpha = \frac{h_E}{\left(\dfrac{V^2}{2g}\right)} = \phi\left[\frac{V_M}{V_L}\right]^2 + \Theta \qquad (1\text{-}68)$$

where V_M = average flow velocity in the manifold upstream from the lateral, ft/s

V_L = average flow velocity in the lateral, ft/s

Values for ϕ and Θ taken from the experimental data presented by Hudson et al. (1979) are given in Table 1-9.

Substituting for α in Eq. (1-67) from Eq. (1-68) yields

$$h_E = \left[\phi\left(\frac{V_M}{V_L}\right)^2 + \Theta\right]\frac{V_L^2}{2g} \qquad (1\text{-}69)$$

Further substitution for h_E in Eq. (1-66) produces the general equation

$$\Delta h = \left[\phi\left(\frac{V_M}{V_L}\right)^2 + \Theta\right]\frac{V_L^2}{2g} + \frac{V_L^2}{2g}$$

or

$$\Delta h = \beta\left(\frac{V_L^2}{2g}\right) \qquad (1\text{-}70)$$

where $\beta = \phi\left(\dfrac{V_M}{V_L}\right)^2 + \Theta + 1.0 \qquad (1\text{-}71)$

TABLE 1-9   *Experimental values for ϕ and Θ for sharp-edged laterals, after Hudson et al. (1979).*

Lateral length	Θ	ϕ
Long*	0.4	0.90
Short†	0.7	1.67

* Long laterals are taken as those whose length is substantially greater than three pipe diameters.

† Short laterals are taken as those whose length is less than three pipe diameters.

Source: Hudson, H. E., Jr., et al., "Dividing-Flow Manifolds with Square-Edged Laterals," *Journal Environmental Engineering Division, ASCE, 105,* 745 (1979). Used with permission of American Society of Civil Engineers.

For perfect flow distribution the head loss through each lateral is the same, hence

$$\frac{\beta_1(V_L)_1^2}{2g} = \frac{\beta_2(V_L)_2^2}{2g} = \ldots = \frac{\beta_i(V_L)_i^2}{2g} = \text{a constant} \qquad \textbf{(1-72)}$$

According to Eq. (1-72), it is possible to write

$$(V_L)_i = (V_L)_1 \sqrt{\frac{\beta_1}{\beta_i}} \qquad \textbf{(1-73)}$$

A flow balance for any manifold system may be expressed as

$$Q_0 = q_1 + q_2 + \ldots + q_i \qquad \textbf{(1-74)}$$

or

$$Q = a_i(V_L)_1 + a_2(V_L)_2 + \ldots + a_i(V_L)_i \qquad \textbf{(1-75)}$$

where Q_0 = total flow into manifold, cfs

q_i = flow in lateral i, cfs

a_i = cross-sectional area of lateral i, ft^2

$(V_L)_i$ = average velocity of flow in lateral i, ft/s

If it is assumed that all the laterals have the same size and shape, Eq. (1-73) may be combined with Eq. (1-75) to give

$$Q_0 = a(V_L)_1 + a(V_L)_1 \sqrt{\frac{\beta_1}{\beta_2}} + a(V_L)_1 \sqrt{\frac{\beta_1}{\beta_3}} + \ldots + a(V_L)_1 \sqrt{\frac{\beta_1}{\beta_i}} \quad \textbf{(1-76)}$$

or

$$Q_0 = a(V_L)_1 \left[1 + \sqrt{\frac{\beta_1}{\beta_2}} + \sqrt{\frac{\beta_1}{\beta_3}} + \ldots + \sqrt{\frac{\beta_1}{\beta_i}} \right] \qquad \textbf{(1-77)}$$

Equation (1-77) may be solved for $(V_L)_1$ so that

$$(V_L)_1 = \frac{Q_0}{a\sqrt{\beta_1}} \left[\sum_{i=1}^{i=n} \sqrt{\frac{1}{\beta_i}} \right]^{-1} \qquad \textbf{(1-78)}$$

where a = cross-sectional area of an individual lateral, ft^2

n = number of laterals

Example Problem 1-9 illustrates how Eq. (1-71), (1-73), and (1-78) are applied in the analysis of dividing-flow manifold problems.

EXAMPLE PROBLEM 1-9: Consider the dividing-flow manifold problem described in Example Problem 1-8.

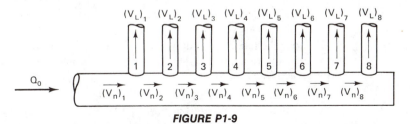

FIGURE P1-9

The total flow into the manifold is 10 million gpd. Determine the flow distribution if a 12-in.-diameter manifold pipe and 4-in.-diameter short laterals are used in the design.

Solution:

1. To solve this problem an iterative procedure is required. It is convenient to construct a computation table for each iteration.

    FIRST ITERATION:

    (a) *Assume* some initial flow distribution to the laterals and enter the assumed values in Column 2 of the first iteration computation table. It is convenient for the first iteration to assume an equal flow distribution to the laterals.

$$q_1 = q_2 = \ldots = q_1 = \frac{1.547 \times 10}{8} = \textbf{1.93 cfs}$$

First iteration computation table for Example Problem 1-9.

Lateral no.	q_i (cfs)	$(V_L)_i$ (fps)	$(Q_M)_i$ (cfs)	$(V_M)_i$ (fps)	$\left[\dfrac{(V_M)_i}{(V_L)_i}\right]^2$	β_i	$\sqrt{\dfrac{1}{\beta_i}}$
(1)	(2)	(3)	(4)	(5)	(6)	(7)	(8)
1	1.93	22.12	15.47	19.70	0.79	3.02	0.58
2	1.93	22.12	13.54	17.25	0.60	2.70	0.61
3	1.93	22.12	11.61	14.79	0.45	2.45	0.64
4	1.93	22.12	9.68	12.33	0.31	2.22	0.67
5	1.93	22.12	7.75	9.87	0.20	2.03	0.70
6	1.93	22.12	5.82	7.41	0.11	1.88	0.73
7	1.93	22.12	3.89	4.96	0.05	1.78	0.75
8	1.93	22.12	1.96	2.50	0.01	1.72	0.76
Total	15.44						5.44

(b) With the flow distribution assumed in Step (a), compute the average flow velocity and enter this value in Column 3 of the first iteration computation table.

$$(V_L)_1 = (V_L)_2 = \ldots = (V_L)_i = \frac{q_i}{a} = \frac{1.93}{\frac{\pi(4/12)^2}{4}} = \textbf{22.12 fps}$$

(c) Considering the reduction in flow at each lateral, compute the flow rate *upstream* from each outlet. Enter the values in Column 4 of the computation table.

$$Q_1 = Q_0 \qquad = 15.47$$
$$Q_2 = Q_1 - q_1 = 13.54$$
$$Q_3 = Q_2 - q_2 = 11.61$$
$$Q_4 = Q_3 - q_3 = 9.68$$
$$Q_5 = Q_4 - q_4 = 7.75$$
$$Q_6 = Q_5 - q_5 = 5.82$$
$$Q_7 = Q_6 - q_6 = 3.89$$
$$Q_8 = Q_7 - q_7 = 1.96$$

(d) Using the data from Column 4, compute the average velocity of flow in the manifold *upstream* from each outlet. Enter the values in Column 5 of the computation table.

$$(V_M)_1 = \frac{Q_1}{A} = \frac{15.47}{0.785} = 19.70$$

$$(V_M)_2 = \frac{Q_2}{A} = \frac{13.54}{0.785} = 17.25$$

$$(V_M)_3 = \frac{Q_3}{A} = \frac{11.61}{0.785} = 14.79$$

$$(V_M)_4 = \frac{Q_4}{A} = \frac{9.68}{0.785} = 12.33$$

$$(V_M)_5 = \frac{Q_5}{A} = \frac{7.75}{0.785} = 9.87$$

$$(V_M)_6 = \frac{Q_6}{A} = \frac{5.82}{0.785} = 7.41$$

$$(V_M)_7 = \frac{Q_7}{A} = \frac{3.89}{0.785} = 4.96$$

$$(V_M)_8 = \frac{Q_8}{A} = \frac{1.96}{0.785} = 2.50$$

(e) Calculate the $[(V_M)_i/(V_L)_i]^2$ term for each lateral and enter the values in the computation table.

$$\left[\frac{(V_M)_1}{(V_L)_1}\right]^2 = \left[\frac{19.70}{22.12}\right]^2 = 0.79$$

$$\left[\frac{(V_M)_2}{(V_L)_2}\right]^2 = \left[\frac{17.25}{22.12}\right]^2 = 0.60$$

$$\left[\frac{(V_M)_3}{(V_L)_3}\right]^2 = \left[\frac{14.79}{22.12}\right]^2 = 0.45$$

$$\left[\frac{(V_M)_4}{(V_L)_4}\right]^2 = \left[\frac{12.33}{22.12}\right]^2 = 0.31$$

$$\left[\frac{(V_M)_5}{(V_L)_5}\right]^2 = \left[\frac{9.87}{22.12}\right]^2 = 0.20$$

$$\left[\frac{(V_M)_6}{(V_L)_6}\right]^2 = \left[\frac{7.41}{22.12}\right]^2 = 0.11$$

$$\left[\frac{(V_M)_7}{(V_L)_7}\right]^2 = \left[\frac{4.96}{22.12}\right]^2 = 0.05$$

$$\left[\frac{(V_M)_8}{(V_L)_8}\right]^2 = \left[\frac{2.50}{22.12}\right]^2 = 0.01$$

(f) Calculate the value for each lateral by applying Eq. (1-71) and using the ϕ and Θ values for short laterals given in Table 1-9. Enter the calculations in Column 7 of the computation table.

$$\beta_1 = \phi\left[\frac{(V_M)_1}{(V_L)_1}\right]^2 + \Theta + 1.0 = (1.67)(0.79) + 0.7 + 1.0 = 3.02$$

$$\beta_2 = (1.67)(0.60) + 0.7 + 1.0 = 2.70$$
$$\beta_3 = (1.67)(0.45) + 0.7 + 1.0 = 2.45$$
$$\beta_4 = (1.67)(0.31) + 0.7 + 1.0 = 2.22$$
$$\beta_5 = (1.67)(0.20) + 0.7 + 1.0 = 2.03$$
$$\beta_6 = (1.67)(0.11) + 0.7 + 1.0 = 1.88$$
$$\beta_7 = (1.67)(0.05) + 0.7 + 1.0 = 1.78$$
$$\beta_8 = (1.67)(0.01) + 0.7 + 1.0 = 1.72$$

(g) Compute the $\sqrt{1/\beta_i}$ term for each lateral and enter the values in the computation table.

$$\sqrt{\frac{1}{\beta_1}} = 0.58 \qquad\qquad \sqrt{\frac{1}{\beta_5}} = 0.70$$

$$\sqrt{\frac{1}{\beta_2}} = 0.61 \qquad\qquad \sqrt{\frac{1}{\beta_6}} = 0.73$$

$$\sqrt{\frac{1}{\beta_3}} = 0.64 \qquad\qquad \sqrt{\frac{1}{\beta_7}} = 0.75$$

$$\sqrt{\frac{1}{\beta_4}} = 0.67 \qquad\qquad \sqrt{\frac{1}{\beta_8}} = 0.76$$

2. Apply Eq. (1-78) and compute the velocity in the first lateral for the second iteration.

$$(V_L)_1 = \frac{Q_0}{a\sqrt{\beta_1}}\left[\sum_{i=1}^{i=n}\sqrt{\frac{1}{\beta_i}}\right]^{-1}$$

$$= \frac{15.47}{(0.087)(1.74)}[5.44]^{-1}$$

$$= 18.81 \text{ fps}$$

3. Using the β values given in the first iteration computation table and the velocity through the first lateral computed in Step 2, estimate the velocity through each of the other laterals and enter the results in the second iteration computation table.

$$(V_L)_i = (V_L)_1\sqrt{\frac{\beta_1}{\beta_i}}$$

$$(V_L)_2 = (18.81)\sqrt{\frac{3.02}{2.70}} = 19.89$$

$$(V_L)_3 = (18.81)\sqrt{\frac{3.02}{2.45}} = 20.88$$

$$(V_L)_4 = (18.81)\sqrt{\frac{3.02}{2.22}} = 21.94$$

$$(V_L)_5 = (18.81)\sqrt{\frac{3.02}{2.03}} = 22.94$$

$$(V_L)_6 = (18.81)\sqrt{\frac{3.02}{1.88}} = 23.84$$

Second iteration computation table for Example Problem 1-9.

Lateral no.	q_i (cfs)	$(V_L)_i$ (fps)	$(Q_M)_i$ (cfs)	$(V_M)_i$ (fps)	$\left[\dfrac{(V_M)_i}{(V_L)_i}\right]^2$	β_i	$\sqrt{\dfrac{1}{\beta_i}}$
(1)	(2)	(3)	(4)	(5)	(6)	(7)	(8)
1	1.64	18.81	15.47	19.71	1.10	3.54	0.53
2	1.73	19.89	13.83	17.62	0.78	3.00	0.58
3	1.82	20.88	12.10	15.41	0.54	2.60	0.62
4	1.91	21.94	10.28	13.09	0.36	2.30	0.66
5	1.99	22.94	8.37	10.66	0.22	2.07	0.70
6	2.07	23.84	6.38	8.13	0.12	1.90	0.73
7	2.13	24.50	4.31	5.26	0.05	1.78	0.75
8	2.17	24.92	2.18	2.78	0.01	1.72	0.76
Total	15.46						5.33

$$(V_L)_7 = (18.81)\sqrt{\frac{3.02}{1.78}} = 24.50$$

$$(V_L)_8 = (18.81)\sqrt{\frac{3.02}{1.72}} = 24.92$$

4. Using the velocity values in Column 3 of the second iteration computation table, complete the calculations required to fill the table.

  (a) Compute the flow in each lateral and enter the values in Column 1 of the computation table.

$$q_i = a(V_L)_i$$
$$q_1 = (0.087)(18.81) = 1.64 \text{ cfs}$$
$$q_2 = (0.087)(19.89) = 1.73 \text{ cfs}$$
$$q_3 = (0.087)(20.88) = 1.82 \text{ cfs}$$
$$q_4 = (0.087)(21.94) = 1.91 \text{ cfs}$$
$$q_5 = (0.087)(22.94) = 1.99 \text{ cfs}$$
$$q_6 = (0.087)(23.84) = 2.07 \text{ cfs}$$
$$q_7 = (0.087)(24.50) = 2.13 \text{ cfs}$$
$$q_8 = (0.087)(24.92) = 2.17 \text{ cfs}$$

  (b) Compute the flow rate upstream from each outlet and enter the values in Column 4 of the computation table.

$$Q_1 = Q_0 \qquad = 15.47$$
$$Q_2 = Q_1 - q_1 = 13.83$$
$$Q_3 = Q_2 - q_2 = 12.10$$
$$Q_4 = Q_3 - q_3 = 10.28$$
$$Q_5 = Q_4 - q_4 = 8.37$$
$$Q_6 = Q_5 - q_5 = 6.38$$
$$Q_7 = Q_6 - q_6 = 4.31$$
$$Q_8 = Q_7 - q_7 = 2.18$$

  (c) Compute the average velocity of flow in the manifold upstream from each outlet and enter the values in Column 5 of the computation table.

$$(V_M)_1 = \frac{Q_1}{A} = \frac{15.47}{0.785} = 19.71$$

$$(V_M)_2 = \frac{Q_2}{A} = \frac{13.83}{0.785} = 17.62$$

$$(V_M)_3 = \frac{Q_3}{A} = \frac{12.10}{0.785} = 15.41$$

$$(V_M)_4 = \frac{Q_4}{A} = \frac{10.28}{0.785} = 13.09$$

$$(V_M)_5 = \frac{Q_5}{A} = \frac{8.37}{0.785} = 10.66$$

$$(V_M)_6 = \frac{Q_6}{A} = \frac{6.38}{0.785} = 8.13$$

$$(V_M)_7 = \frac{Q_7}{A} = \frac{4.31}{0.785} = 5.26$$

$$(V_M)_8 = \frac{Q_8}{A} = \frac{2.18}{0.785} = 2.78$$

(d)   Calculate the $[(V_M)_i/(V_L)_i]^2$ term for each lateral and enter the values in the computation table.

$$\left[\frac{(V_M)_1}{(V_L)_1}\right]^2 = 1.10$$

$$\left[\frac{(V_M)_2}{(V_L)_2}\right]^2 = 0.78$$

$$\left[\frac{(V_M)_3}{(V_L)_3}\right]^2 = 0.54$$

$$\left[\frac{(V_M)_4}{(V_L)_4}\right]^2 = 0.36$$

$$\left[\frac{(V_M)_5}{(V_L)_5}\right]^2 = 0.22$$

$$\left[\frac{(V_M)_6}{(V_L)_6}\right]^2 = 0.12$$

$$\left[\frac{(V_M)_7}{(V_L)_7}\right]^2 = 0.05$$

$$\left[\frac{(V_M)_8}{(V_L)_8}\right]^2 = 0.01$$

(e)   Determine the β value for each lateral and enter the values in the computation table.

$$\beta_1 = \phi\left[\frac{(V_M)_1}{(V_L)_1}\right] + \Theta + 1.0$$

$$= (1.67)(1.1)\ + 0.7 + 1.0 = 3.54$$
$$\beta_2 = (1.67)(0.78) + 0.7 + 1.0 = 3.00$$
$$\beta_3 = (1.67)(0.54) + 0.7 + 1.0 = 2.60$$
$$\beta_4 = (1.67)(0.36) + 0.7 + 1.0 = 2.30$$
$$\beta_5 = (1.67)(0.22) + 0.7 + 1.0 = 2.07$$
$$\beta_6 = (1.67)(0.12) + 0.7 + 1.0 = 1.90$$
$$\beta_7 = (1.67)(0.05) + 0.7 + 1.0 = 1.78$$
$$\beta_8 = (1.67)(0.01) + 0.7 + 1.0 = 1.72$$

(f)   Compute the $\sqrt{1/\beta_i}$ term for each lateral and enter the values in the computation table.

$$\sqrt{\frac{1}{\beta_1}} = 0.53$$

$$\sqrt{\frac{1}{\beta_2}} = 0.58$$

$$\sqrt{\frac{1}{\beta_3}} = 0.62$$

$$\sqrt{\frac{1}{\beta_4}} = 0.66$$

$$\sqrt{\frac{1}{\beta_5}} = 0.70$$

$$\sqrt{\frac{1}{\beta_6}} = 0.73$$

$$\sqrt{\frac{1}{\beta_7}} = 0.75$$

$$\sqrt{\frac{1}{\beta_8}} = 0.76$$

The iterative calculation procedure is continued until the difference between the lateral flows for the last two iterations agree within some specified error criterion. For this problem five iterations are required to satisfy a relative difference criterion of 1×10^{-2}. The lateral flows after five iterations are

$$(q_L)_1 = 1.486 \text{ cfs}$$
$$(q_L)_2 = 1.648 \text{ cfs}$$
$$(q_L)_3 = 1.798 \text{ cfs}$$
$$(q_L)_4 = 1.930 \text{ cfs}$$
$$(q_L)_5 = 2.042 \text{ cfs}$$
$$(q_L)_6 = 2.132 \text{ cfs}$$
$$(q_L)_7 = 2.197 \text{ cfs}$$
$$(q_L)_8 = 2.237 \text{ cfs}$$

These values show that 9.6% of the total flow passes through the first lateral, while 14.5% passes through the last lateral.

Note: The flow distribution computed for this problem ($q_N/q_1 = 2.237/1.486 = 1.5$) is the same assumed in Example Problem 1-8. However, in the latter problem a much larger lateral diameter (7.7 in.) was estimated.

See Appendix I for the Fortran computer code for this problem.

The accuracy of the procedure outlined in Example Problem 1-9 is probably much greater than that for the procedure outlined in Example Problem 1-8. The iterative technique is useful as both an analysis and a design tool. When used for design, the best approach is to write a short program and simply screen a number of different combinations of lateral and manifold diameters. When a combination that gives an acceptable flow distribution is found, the values are used for design.

REFERENCES

AMIRTHARAJAH, A., "Design of Granular-Media Filter Units," In *Water Treatment Plant Design,* edited by R. L. Sanks, Ann Arbor Science Publishers, Inc., Ann Arbor, Michigan (1978).

BRATER, E. F., and KING, H. W., *Handbook of Hydraulics,* 6th Edition, McGraw–Hill Book Company, New York, New York (1976).

Brisbin, S. G., "Flow of Concentrated Raw Sewage Sludges in Pipes," *Journal Sanitary Engineering Division, ASCE,* **83**, 201, (1957).

Camp, T. R., "Applied Hydraulic Design of Treatment Plants, (Part I)," In *Seminar Papers on Wastewater Treatment and Disposal,* edited by G. M. Reece, Boston Society of Civil Engineers, 231, Boston, Massachusetts (1961).

Fair, G. M., "The Hydraulics of Rapid Sand Filters," *Journal Institute of Water Engineers,* **5,** 171 (1951).

Hudson, H. E., Jr., Uhler, R. B., and Bailey, R. W., "Dividing-Flow Manifolds with Square-Edged Laterals," *Journal Environmental Engineering Division, ASCE,* **105,** 745 (1979).

Jain,, A. K., Mohan, D. M., and Khanna, P., "Modified Hazen–Williams Formula," *Journal Environmental Engineering Division, ASCE,* **104,** 137 (1978).

Jeppson, R. W., *Analysis of Flow in Pipe Networks,* Ann Arbor Science Publishers, Inc., Ann Arbor, Michigan (1977).

Metcalf and Eddy, Inc., *Wastewater Engineering,* 1st Edition, McGraw–Hill Book Company, New York, New York (1972).

Rich, L. G., *Unit Operations of Sanitary Engineering,* John Wiley & Sons, Inc., New York, New York (1961).

Simon, A. L., *Practical Hydraulics,* John Wiley & Sons, Inc., New York, New York (1976).

Vesilind, P. A., *Treatment and Disposal of Wastewater Sludges,* Ann Arbor Science Publishers, Inc., Ann Arbor, Michigan (1979).

Wood, D. J., "An Explicit Friction Factor Relationship," *Civil Engineering, ASCE,* **36,** 60 (December 1966).

2

MULTIPORT DIFFUSER OUTFALLS

The disposal of wastewater from treatment plants is often accomplished by discharging the effluents into lakes, rivers, estuaries, and oceans where the wastes are further decomposed by natural processes. Although discharge canals and pipelines are sometimes used to introduce wastewater effluents into natural water bodies as "point sources," multiport diffusers discharge over larger areas, and thus, produce greater dilution for the same effluent and ambient conditions.

A multiport diffuser is essentially a large manifold pipe equipped with a series of outlet ports along its length. The effluent is discharged through the outlets as high-velocity buoyant jets that entrain large volumes of ambient water as they issue from the manifold and rise toward the water surface (see Fig. 2-1). The manifold pipe is located on or beneath the bottom of the water body, so that the length of the trajectories of the rising buoyant jets is maximized. This ensures that the greatest possible initial dilution is realized. Sections of two idealized multiport diffusers are shown in Fig. 2-2.

Usually, a multiport diffuser is designed to discharge the effluent uniformly along the length of the diffuser. However, for water bodies with nonuniform ambient discharge distributions along the diffuser, the best mixing is achieved when the

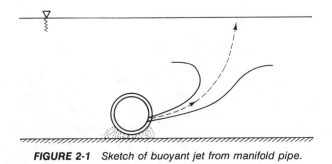

FIGURE 2-1 *Sketch of buoyant jet from manifold pipe.*

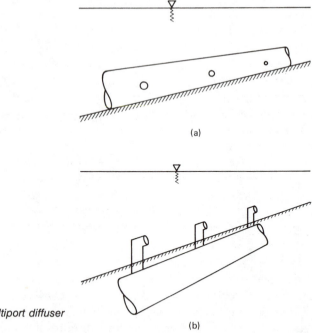

FIGURE 2-2 *Idealized multiport diffuser pipe sections.*

effluent is discharged approximately proportional to the ambient discharge distribution at the outfall section. This approach is recommended for multiport diffuser installations in rivers.

This chapter will consider the internal hydraulics and design of multiport diffusers. Specific design procedures based on theoretical and empirical curves and equations are presented for multiport diffusers in rivers and oceans.

2.1 HYDRAULICS OF MULTIPORT DIFFUSERS

The flow in the neighborhood of an individual diffuser port may be considered independent of the rest of the diffuser flow provided the port spacing is greater than about 10 port diameters. When these conditions are met, the discharge from a diffuser port can be given by

$$q = C_D a \sqrt{2gE} \tag{2-1}$$

in which C_D = discharge coefficient; $a = \pi d^2/4$ = port area; g = gravitational acceleration; and E = difference in total head across the port; that is, the difference in pressure head across the port plus the velocity head of the flow in the manifold pipe.

When the port discharge, q, is much smaller than the flow in the diffuser and the port diameter is less than $1/4$ of the manifold diameter, the theoretical analysis of branching flow by McKnown and Hsu (1951) can be extended to express the discharge coefficient, C_D, as a function of the ratio of the velocity head in the manifold to the total energy. The empirical relationships for sharp-edged and bell-mouthed ports are

$$C_D = 0.63 - 0.58 \left[\frac{V^2}{2gE} \right] \qquad (2\text{-}2)$$

and

$$C_D = 0.975 \left[1 - \frac{V^2}{2gE} \right]^{0.375} \qquad (2\text{-}3)$$

respectively, where V is the manifold pipe velocity at the section being considered. For geometries or conditions not matching those assumed for Eqs. (2-2) and (2-3), experimental studies must be performed in order to develop unique empirical relationships for the discharge coefficient, C_D. Buried manifold pipes, for instance, require special riser assemblies like those shown in Fig. 2-2(b). The discharge coefficient for this type of outlet will generally be dependent on several geometric and flow variables. Fischer et al. (1979) present a discussion of discharge coefficients for riser-type ports.

The fundamental studies of McKnown (1954) and McKnown and Hsu (1951) have shown that it is reasonable to assume no energy loss for the manifold flow in passing a discharge port. That is, it can be assumed that perfect pressure recovery compensates for the decrease in velocity of the main flow across a lateral discharge section. Therefore, the difference in total energy, E, between ports along the diffuser can be accounted for by considering the friction loss due to the main flow using the Darcy–Weisbach equation and by considering the difference in pressure head due to density differences between the effluent and ambient fluids for sloping pipes. When the density of the ambient and effluent water is the same, the pressure differential across a port is independent of elevation, and only frictional losses need be considered. For sloped diffusers in ocean environments, however, the density difference may be important, since the pressure differential across a port is dependent on the port elevation. As a simplified example of this effect, consider the section of diffuser pipe shown in Fig. 2-3 with very small ports located ℓ and $\ell + \Delta z$ below the saltwater

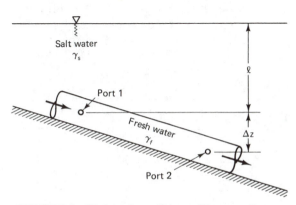

FIGURE 2-3 *Sketch of an offshore diffuser section on a sloped bottom.*

surface. The specific weights of the fresh water and the salt water are denoted by γ_f and γ_s, respectively. If frictional losses are neglected in the manifold pipe, the pressure differences across port 1 and port 2 are given by

$$\Delta P_1 = P_1 - \gamma_s \ell \tag{2-4}$$

$$\Delta P_2 = P_1 + \Delta z \gamma_f - \gamma_s(\ell + \Delta z) = \Delta P_1 - \Delta z(\gamma_s - \gamma_f) \tag{2-5}$$

where P_1 is the pressure within the pipe at port 1. The difference in head due to the density difference is, therefore,

$$\Delta E = \frac{\Delta P_1}{\gamma_f} - \frac{\Delta P_2}{\gamma_f} = \Delta z \left(\frac{\gamma_s - \gamma_f}{\gamma_f} \right)$$

Consequently, density difference tends to produce smaller discharge for deeper ports.

An iterative procedure developed by Rawn et al. (1961) provides the most accurate method of calculating the discharge and head distribution for multiport diffuser design. Essentially, the method entails assuming a value of the total energy, E_1, at the downstream diffuser port, then proceeding step by step toward the upstream end of the diffuser, accounting for changes in energy due to friction and elevation differences and for changes in pipe discharge at lateral outlet sections. The procedure is repeated with new values for E_1, until the specified total discharge or head at the upstream end of the diffuser is obtained. If the total discharge, upstream head, or discharge distribution predicted by the iterative procedure is not satisfactory, it is necessary to modify port or manifold sizes or port spacings and repeat the iterative process until a suitable design is obtained. Although this method seems time consuming, it is easily carried out on a digital computer or even on a hand-held programmable calculator.

The calculation procedure described by Rawn et al. (1961) can best be explained by considering the definition sketch of a multiport diffuser shown in Fig. 2-4. The diffuser ports are numbered from 1 to N starting from the offshore (downstream) end of the diffuser. Note that the manifold diameter, D, the main velocity, V, the port spacing, L, and the port elevation, z, can vary along the diffuser. The elevation, the port diameter, and the total energy corresponding to the ith port from the downstream end of the diffuser can be represented by z_i, d_i, and E_i, respectively. Also, the frictional

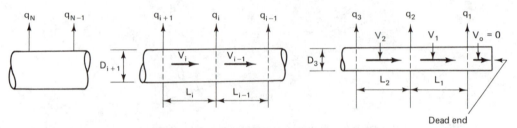

FIGURE 2-4   Definition sketch of a multiport diffuser.

head loss for the manifold flow from the $(i + 1)$th to the ith port is denoted by h_{f_i}. After assuming the total energy, E_1, at the first port (the port nearest the downstream end of diffuser), the calculations proceed upstream from $i = 1$ to $i = N$ (the total number of ports) in the following order:

$$C_{D_i} = \text{a function of } (V_{i-1})^2/2gE_i, \text{ where } V_o = 0 \qquad (2\text{-}6)$$

$$q_i = C_{D_i} \frac{\pi d_i^2}{4} \sqrt{2gE_i} \qquad (2\text{-}7)$$

$$Q_i = \frac{4}{\pi D_i^2} \left[\sum_{i=1}^{i=N} q_i \right] \qquad (2\text{-}8)$$

$$V_i = \frac{4 \left[\sum_{i=1}^{i=N} q_i \right]}{\pi D_i^2} \qquad (2\text{-}9)$$

$$h_{f_i} = f \frac{L_i}{D_i} \frac{V_i^2}{2g} \qquad (2\text{-}10)$$

$$E_{i+1} = E_i + h_{f_i} + \left[\frac{\gamma_r - \gamma_e}{\gamma_e} \right] (z_{i+1} - z_i) \qquad (2\text{-}11)$$

where γ_r = specific weight of receiving water and γ_e = specific weight of wastewater effluent. The Darcy–Weisbach friction factor, f, can be assumed constant or it can be calculated more accurately at each step by iteration from the Colebrook–White equation:

$$\frac{1}{\sqrt{f}} - 2 \log \left[\frac{D}{2e} \right] = 1.74 - 2 \log \left[1 + 18.7 \frac{\left(\frac{D}{2e} \right)}{\text{Re} \sqrt{f}} \right] \qquad (2\text{-}12)$$

in which e is the equivalent sand roughness of the manifold pipe and $\text{Re} = VD/\nu$ is the Reynolds number of the pipe flow where ν is the kinematic viscosity of the sewage effluent. If the total head, E_N, and manifold discharge, $Q_N = V_N[\pi D_N^2/4]$ at the Nth port do not agree with the available head at the upstream end of the diffuser or the desired total discharge (whichever is specified), a new value of E_1 is assumed and the procedure is repeated.

It is interesting to note that for diffuser installations where elevation changes are not important, only one iteration is necessary, since all head terms are proportional to the discharge terms squared. For example, if the first iteration results in $E_N = E_N'$ and $Q_N = Q_N'$, and the specified (or available) head at the upstream port is E_s, the corrected head and discharge for the ith port are given by

$$E_i = \left[\frac{E_s}{E_N'} \right] E_i' \qquad (2\text{-}13)$$

and

$$q_i = \left[\frac{E_s}{E_N'}\right]^2 q_i' \tag{2-14}$$

in which E_i' and q_i' are the total energy and the port discharge, respectively, determined for the ith port by the iteration procedure. If, on the other hand, the total discharge is specified as Q_s and the iteration results in $E_N = E_N'$ and $Q_N = Q_N'$, the correct values for the total head and the port discharge for the ith port are

$$E_i = \left[\frac{Q_s}{Q_N'}\right]^{1/2} E_i' \tag{2-15}$$

and

$$q_i = \left[\frac{Q_s}{Q_N'}\right] q_i' \tag{2-16}$$

When both available head and total discharge are specified, it will usually be necessary, even for freshwater diffusers, to examine various manifold and/or port diameters in order to meet both criteria.

Several analyses simpler than those of Rawn et al. have been developed for multiport diffusers when density differences have negligible effect on the diffuser hydraulics. French (1972) stated that this condition is met when

$$\left[\frac{\gamma_r - \gamma_e}{\gamma_e}\right] \Delta z_{\max} \ll E_N \tag{2-17}$$

where $\Delta z_{\max}$ = difference in head between the highest and the lowest diffuser ports.

The most well known analysis for the condition specified by Eq. (2-17) is the continuous model of Camp and Graber (1968). By assuming a uniformly decreasing discharge along the length of a constant-diameter diffuser manifold and a constant-friction factor, they developed the following equation for the change in piezometric head, Δh, from the diffuser inlet downstream, a distance, x, from the inlet

$$\Delta h_x = h - h_o = \frac{V_o^2}{2g}\left\{1 - \left(1 - \frac{x}{L}\right)^2 - \frac{FL}{3}\left[1 - \left(1 - \frac{x}{L}\right)^3\right]\right\} \tag{2-18}$$

in which h = piezometric head at downstream distance, x, from inlet; h_o = piezometric head at inlet; V_o = velocity at inlet to diffuser manifold; L = length of diffuser; and $F = f/D$ = friction factor divided by pipe diameter. Equation (2-18) can be used to determine the distribution of piezometric head along the diffuser, which can, in turn, be used to determine the discharge distribution for a proposed design using the discharge equation, Eq. (1-31),

$$q = C_d a \sqrt{2g\Delta h}$$

where C_d is assumed constant. The relationship between the discharge coefficient in this equation and the coefficient, C_D, in Eq. (2-1) is

$$C_d = \frac{C_D}{\sqrt{1 - \dfrac{V^2}{2gE}}} \qquad (2\text{-}19)$$

Camp and Graber's method was extended by Jain et al. (1971) in the development of a model that describes the piezometric head and discharge distributions along a diffuser pipe consisting of a series of outfall sections with different outflow characteristics. This type of diffuser is used when it is desirable to discharge the effluent in approximate proportion to the ambient discharge distribution in a river. This results in very large initial dilutions. A schematic depicting a multiport diffuser consisting of N sections, each having different outflow characteristics is shown in Fig. 2-5. The effluent discharge is assumed to be uniform along any given section.

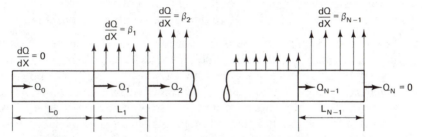

FIGURE 2-5 *Schematic of a multiport diffuser consisting of* N *sections with different outflow characteristics, after Jain et al. (1971).*

The difference between the total head at the upstream end of the diffuser, H_o, and the piezometric head at the nth junction is given by

$$H_o - h_n = \frac{Q_o^2}{2gA^2} - \sum_{i=1}^{n-1} \frac{Q_i^2}{2gA^2}\left[2k_i - k_i^2 - \frac{fL_i}{D}\left(1 - k_i + \frac{k_i^2}{3}\right)\right] \qquad (2\text{-}20)$$

in which $k_i = (Q_i - Q_{i-1})/Q_i$. After Eq. (2-20) is used to determine the head distribution along the diffuser for specified values of pipe diameter and outflow for each section, Eq. (1-31) can be solved for the size and number of ports for each outflow section.

Although these other methods are interesting and useful for specific diffuser conditions, the discrete model of Rawn et al. (1961) is generally used on the final design of sewage diffusers. Therefore, three of the four design examples considered herein will use only the discrete model. The last example employs the continuous model of Camp and Graber (1968).

2.2 DESIGN OF MULTIPORT DIFFUSERS

The design of multiport diffusers for discharging sewage effluents into natural water bodies includes the following hydraulic requirements:

1. The outflow distribution should be nearly uniform along the diffuser for lake and ocean installations where ambient velocities are zero or nearly constant along the diffuser. In rivers, where significant transverse variations in ambient velocity are typical, the discharge of sewage effluent should be approximately proportional to the ambient discharge distribution, if initial dilution is to be maximized.

2. Velocities in the manifold pipe should be at least 2 to 3 ft/sec in order to inhibit the accumulation of settleable solids in the diffuser.

3. If lower velocities must be used, provisions should be made to facilitate periodic cleaning or flushing of grease, slimes, and grit from the diffuser. This can be accomplished by installing a removable bulkhead that provides access through the downstream end of the diffuser. It may also be advisable to place an orifice in the bulkhead to provide some "continuous flushing."

4. In order to prohibit seawater intrusion for offshore diffusers, all ports must flow full. This is ensured when the port velocity is greater than the manifold velocity at all sections along the diffuser. Rawn et al. (1961) noted that this condition is met when the sum of the effective area of all ports downstream from any section does not exceed the cross-sectional area of the manifold pipe at the section. In practice, ratios of total port area to pipe area between $1/3$ and $2/3$ produce the best diffuser performance (Fischer et al., 1979).

5. The frictional head loss should be minimized, within the constraints of the other requirements, in order to minimize pumping costs.

6. Ports should be oriented horizontally and, if possible, designed with smooth corners so that material does not accumulate and reduce the effective area of the port. Jet interference is reduced and, thus, initial dilution is increased when ports are located alternately from side to side along the manifold pipe. Typical ranges for port diameters and spacings are 3 to 9 in. and 8 to 15 ft, respectively.

The application of the discrete model summarized by Eqs. (2-6) through (2-11), subject to the requirements discussed above, will be considered in design Example Problems 2-1, 2-2, and 2-3.

EXAMPLE PROBLEM 2-1: Design an ocean outfall diffuser to discharge average and peak wastewater flows of 20 and 35 cfs, respectively, for a maximum available head at the treatment plant of 23 ft. The diffuser, located $1/2$ mile offshore where the bottom slope is 0.030, is to be 200 ft long and have sharp-edged ports spaced every 10 ft. The manifold pipe has an equivalent sand roughness of 0.005 ft. The specific weight of salt water and wastewater may be taken as 64.0 and 62.4 lb/ft^3, respectively.

Solution:

1. Establish the size of the manifold pipe. In order to prevent excessive buildup of settleable solids in the manifold pipe, the velocity at the middle of the pipe will be maintained at about 2 fps for the average flow rate (Q) of 20 cfs. These criteria can be used to calculate the diameter of the manifold pipe as follows:

$$Q_{mid} = V_{mid}A = V_{mid}\left[\frac{\pi D^2}{4}\right]$$

therefore, assuming a mid-pipe flow rate of 10 cfs,

$$D = \sqrt{\frac{4Q_{mid}}{\pi V_{mid}}} = \sqrt{\frac{4(10 \text{ cfs})}{\pi(2 \text{ fps})}} = 2.52 \text{ ft}$$

or, rounding to a standard size, $D = $ **2.5 ft.**

2. Compute the energy at the Nth port, E_N, for the peak flow condition. This can be determined by subtracting the head loss in the $1/2$-mile pipeline from the available head, H, at the plant. If the pipeline is also 2.5 ft in diameter, the friction factor, f, is .024 for $e/D = 0.005/2.5 = 0.002$. Therefore, neglecting minor losses, E_N, is determined from the energy equation as follows:

$$E_N = H - \frac{8fL_oQ_p^2}{\pi^2 gD^2} = 23 - \frac{8(0.024)(2640)(35)^2}{\pi^2(32.2)(2.5)^5}$$

where Q_p represents the peak wastewater flow rate.

$$E_N = \textbf{2.99 ft}$$

where $L_o = $ distance from plant to manifold pipe, ft

3. Determine the required port diameter. The port diameter, d, must be less than $1/4$ the pipe diameter and less than $1/10$ of the port spacing in order for the discrete model of Rawn et al. (1961) to be applicable. The first restriction imposes the governing limit $d < 7.5$ in. for this design. An initial estimate of d can be made at the Nth port from Eq. (2-1) as follows:

$$d_N \simeq \sqrt{\frac{4q}{C_{D_N}\pi[2gE_N]^{1/2}}}$$

where $q = \dfrac{Q_p}{N} = \dfrac{35 \text{ cfs/diffuser pipe}}{20 \text{ ports/diffuser pipe}} = 1.75 \text{ cfs/port}$

and

$$C_{D_N} = 0.63 - 0.58\left[\frac{Q_p^2}{2g\left[\dfrac{\pi D^2}{4}\right]^2 E_N}\right] = 0.477$$

Thus,

$$d_N \simeq \sqrt{\frac{4(1.75)}{(0.477)\pi[(2)(32.2)(2.99)]^{1/2}}} = 0.58 \text{ ft} = \textbf{6.96 in.}$$

This estimate, rounded to 7 in., can be used as an initial estimate for all of the port diameters.

4. Apply the model of Rawn et al. to determine by iteration the port diameter that will allow the peak discharge to pass through the diffuser at the available head (the required computer code is given in Appendix I).

The first set of iterations using $d = 7$ in. is summarized in Table 2-1. Essentially, the procedure entails finding the value of the head at the downstream port, E_1, that gives the desired discharge, then checking to see if the head at the Nth port, E_N, is equal to the specified value determined in Step 2. For this case, $Q_p \simeq 35$ cfs for $E_1 = 1.903$ ft and $E_N = 2.541$ ft. Table 2-1 shows that if the diffuser were equipped with 7-in. diameter ports it would, in fact, pass 35 cfs at less than 2.99 ft, the available head at the Nth port. It is, however, desirable to maximize jet velocities from the ports in order to promote efficient mixing. Consequently, the smallest port diameter that will allow the peak flow to pass at the available head should be chosen.

The port discharges are shown in Table 2-2 for the conditions specified in the last row of Table 2-1. The discharge distribution varies along the diffuser by as much as 17%. This is usually not acceptable. In this case two different port sizes will be needed; d_1 for ports 1–10 and d_2 for ports 11–20.

Proportional estimates for d_1 and d_2 can be made by noting from Eq. (2-1) that $d \propto q^{1/2} E^{-1/4}$. Therefore, using values from Table 2-2 for q near the middle of the two 10-port sections of the diffuser

$$d_1 \simeq (7 \text{ in.}) \left[\frac{1.750}{1.838}\right]^{1/2} \left[\frac{2.993}{2.541}\right]^{-1/4} = 6.56 \text{ in.}$$

$$d_2 \simeq (7 \text{ in.}) \left[\frac{1.750}{1.670}\right]^{1/2} \left[\frac{2.993}{2.541}\right]^{-1/4} = 6.88 \text{ in.}$$

TABLE 2-1 Values of E_N and Q_p obtained using Rawn et al.'s method (Example Problem 2-1).

D (ft)	d (in.)	E_1 (ft)	E_N (ft)	Q_p (cfs)
(1)	(2)	(3)	(4)	(5)
2.5	7.0	2.000	2.663	35.844
2.5	7.0	1.900	2.538	34.971
2.5	7.0	1.903	2.541	34.997

TABLE 2-2 *Distributions of port discharge manifold pipe velocity, and head for d = 7 in., Q_p = 35 cfs, E_1 = 1.903 ft, and E_N = 2.541 ft (Example Problem 2-1).*

Port no.	Port discharge (cfs)	Pipe velocity (fps)	Head (ft)
(1)	*(2)*	*(3)*	*(4)*
1	1.863	0.380	1.903
2	1.865	0.760	1.911
3	1.863	1.139	1.919
4	1.858	1.518	1.929
5	1.850	1.895	1.940
6	1.838	2.270	1.953
7	1.824	2.641	1.968
8	1.808	3.010	1.987
9	1.791	3.375	2.008
10	1.772	3.736	2.032
11	1.752	4.093	2.061
12	1.731	4.445	2.094
13	1.710	4.794	2.131
14	1.690	5.138	2.173
15	1.670	5.479	2.220
16	1.652	5.815	2.272
17	1.634	6.148	2.330
18	1.618	6.478	2.394
19	1.604	6.805	2.465
20	1.592	7.129	2.541

TABLE 2-3 *Values of E_N and Q_p obtained using Rawn et al.'s method (Example Problem 2-1).*

D (ft)	d_1 (in.)	d_2 (in.)	E_1 (ft)	E_N (ft)	Q_p (cfs)
(1)	*(2)*	*(3)*	*(4)*	*(5)*	*(6)*
2.5	6.56	6.88	2.200	2.817	34.986
2.5	6.56	6.88	2.210	2.830	35.062
2.5	6.56	6.88	2.202	2.820	35.001

The results of the iterations for these values of d_1 and d_2 are summarized in Table 2-3. The discharge distribution for the conditions of the last line in Table 2-3 is given in Table 2-4. The variation in port discharge has been decreased to a maximum of about 10% which is acceptable for offshore diffusers. The value of E_N in Table 2-3 is still below the specified head of 2.99 ft. Consequently, smaller ports could be used.

Adopting standard sized ports of d_1 = 6.5 in. and d_2 = 6.75 in., results in Q_p = 35.000 for E_1 = 2.319 ft and E_N = 2.939 ft. Although E_N is still less than the

TABLE 2-4 Distributions of port discharge and
manifold pipe velocity for d_1 = 6.56 in., d_2 =
6.88 in., Q_p = 35 cfs, E_1 = 2.202 ft, and E_N =
2.820 ft (Example Problem 2-1).

Port no.	Port discharge (cfs)	Pipe velocity (fps)
(1)	(2)	(3)
1	1.761	0.359
2	1.763	0.718
3	1.762	1.077
4	1.758	1.435
5	1.752	1.792
6	1.744	2.147
7	1.734	2.501
8	1.723	2.852
9	1.710	3.200
10	1.696	3.545
11	1.848	3.922
12	1.828	4.294
13	1.808	4.663
14	1.787	5.027
15	1.767	5.387
16	1.747	5.743
17	1.728	6.095
18	1.710	6.443
19	1.694	6.788
20	1.680	7.130

specified value, it is close enough. The port discharge and pipe velocity distributions for this final design are shown in Table 2-5. The total discharges for the first 10 ports and for the last 10 ports are 17.54 and 17.46 cfs, respectively. The port discharge and the pipe velocity distributions for the average flow of 20 cfs are also shown in Table 2-5. The total discharges for the first 10 ports and for the last 10 ports are 9.83 and 10.17 cfs, respectively, for Q = 20 cfs.

EXAMPLE PROBLEM 2-2: Wastewater is to be discharged into a river from a treatment plant at average and peak flow rates of 9 and 20 cfs, respectively. The river is 520 ft wide at the discharge section and the ambient discharge distribution is as shown in Fig. 2-6. Design an effective multiport diffuser subject to the following conditions:

1. Available head at the plant = 6 ft

2. f = 0.025 for manifold pipe

3. Sharp-edged orifices

TABLE 2-5 *Distribution of port discharge and manifold pipe velocity for final design conditions: D = 2.5 ft, d_1 = 6.5 in., and d_2 = 6.75 in. (Example Problem 2-1).*

Port no.	Pipe diameter (ft)	Port diameter (in.)	Port discharge for Q_p = 35 cfs (cfs)	Pipe velocity for Q_p = 35 cfs (fps)	Port discharge for Q = 20 cfs (cfs)	Pipe velocity for Q = 20 cfs (fps)
(1)	(2)	(3)	(4)	(5)	(6)	(7)
1	2.5		1.774	0.361	0.978	0.199
2	2.5		1.776	0.723	0.983	0.400
3	2.5		1.775	1.085	0.986	0.600
4	2.5		1.770	1.446	0.988	0.802
5	2.5	6 1/2	1.766	1.805	0.989	1.003
6	2.5		1.758	2.163	0.987	1.204
7	2.5		1.748	2.520	0.986	1.405
8	2.5		1.737	2.873	0.983	1.605
9	2.5		1.724	3.225	0.980	1.805
10	2.5		1.710	3.573	0.976	2.003
11	2.5		1.829	3.946	1.047	2.217
12	2.5		1.810	4.314	1.041	2.429
13	2.5		1.791	4.679	1.033	2.639
14	2.5		1.772	5.040	1.026	2.848
15	2.5	6 3/4	1.753	5.397	1.019	3.056
16	2.5		1.734	5.750	1.012	3.262
17	2.5		1.717	6.100	1.005	3.467
18	2.5		1.700	6.446	0.999	3.670
19	2.5		1.685	6.790	0.994	3.873
20	2.5		1.671	7.130	0.989	4.074

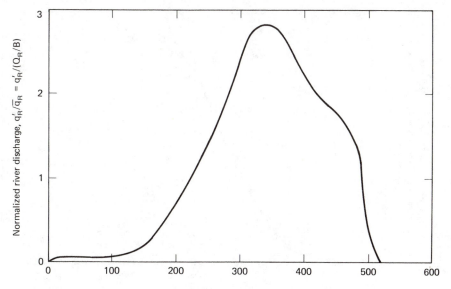

FIGURE 2-6 *Normalized river discharge distribution at cross section. (Note: q_R = uy = local discharge per unit width, Q_R = total river discharge, and B = river width.)*

Solution:

1. Establish the effluent discharge distribution that approximates the river discharge distribution at the outfall section. This can be accomplished by approximately matching the river discharge distribution with a stepwise effluent discharge distribution over the region of the river width in which the wastewater will be discharged. This is exemplified for the example considered herein in Fig. 2-7. The dashed lines represent the stepwise diffuser discharge for $z = 180$ to 480 ft. The wastewater discharge of each of the diffuser sections will be proportional to the ordinates, $q_e/\overline{q}_e = 1.91, 2.67$, and 1.15, read from Fig. 2-7 for Secs. 1, 2, and 3, respectively. Therefore, the wastewater discharges from Secs. 1, 2, and 3 for average operating conditions are

$$Q_1 = \textbf{3.00 cfs}$$

$$Q_2 = \textbf{4.19 cfs}$$

$$Q_3 = \textbf{1.81 cfs}$$

respectively. Similarly, the respective discharges for peak wastewater flow are

$$Q_{1_p} = \textbf{6.67 cfs}$$

$$Q_{2_p} = \textbf{9.32 cfs}$$

$$Q_{3_p} = \textbf{4.01 cfs}$$

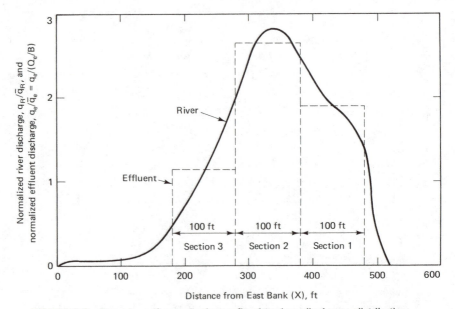

FIGURE 2-7 *Stepwise effluent discharge fitted to river discharge distribution.* *(Note: q_e = local effluent discharge per unit width, Q_e = total effluent discharge.)*

2. Compute the diameter of the manifold pipe. This pipe must be small enough so that the velocity will inhibit the buildup of solids within the pipe for average flow conditions. In this case, if the pipe velocity at the downstream end of Sec. 2 is equal to 2 fps for average flow, the diameter of the manifold pipe is

$$D = \frac{4(Q_3 + Q_2)}{\pi V} = \frac{4(6 \text{ cfs})}{\pi(2 \text{ fps})} = 1.95 \text{ ft}$$

   or, rounding to standard units, let $D = $ **2 ft.**

3. Determine the total head at the upstream end of the outflow portion of the diffuser. The remainder of the hydraulic design, in which the size and spacing of the ports are determined, is based on the peak discharge. Thus, total head, E_N, at the upstream end of the outflow portion of the diffuser for the peak discharge of 20 cfs, is determined by subtracting the friction head loss in the 180-ft section of pipe without ports from the available head as follows
   $E_N = H - 8fLQ_p^2/\pi^2 gD^5 = 6 - 8(.025)(180 \text{ ft})(20 \text{ cfs})^2/\pi^2 g(2 \text{ ft})^5$

$$E_N = \textbf{4.584 ft}$$

4. Select port sizes and spacings so that the desired discharge distribution is obtained at peak flow for the available head. The procedure involves choosing a port spacing, then using the discrete model to determine the appropriate pipe diameters for each of the sections by trial and error. This mathematical model is valid when $L \geq 10\, d$ and $d \leq D/4$. It may not be possible to satisfy these restrictions for the assumed port spacing at the specified peak discharge and available head, however, and a smaller value of port spacing must be adopted if the design is to be based on the discrete model of Rawn et al. (1961). For this example, a port spacing of 10 ft will be considered initially. Each of the three 100-ft diffusing sections will have 10 ports for a 10-ft port spacing. The diameter of the Nth diffuser port (the upstream port) can be estimated from Eq. (2-1) as follows:

$$d_N \simeq \sqrt{\frac{4q_3}{C_{D_N}\pi[2gE_N]^{1/2}}}$$

where $q_3 = Q_3/10 = 0.401$ cfs = average individual port discharge in Sec. 3.

$$C_{D_N} = 0.63 - 0.58 \frac{V_N^2}{2gE_N} = 0.63 - 0.58 \frac{Q_{\text{peak}}^2}{2g\left(\dfrac{\pi D^2}{4}\right)^2 E_N}$$

$$C_{D_N} = 0.63 - \frac{0.58(20)^2}{2g[\pi 2^2/4]^2} = 0.550$$

Therefore

$$d_N \simeq \sqrt{\frac{4(0.401)}{0.55\pi[2g(4.584)]^{1/2}}} = .232 \text{ ft} = 2.79 \text{ in.}$$

This diameter will be assumed for all the ports in Sec. 3.

Rough initial estimates of the port diameters for the remaining sections can be obtained by noting from Eq. (2-1) that port diameter is proportional to $q^{1/2}$. Thus, by assuming that $E = E_N$ for the entire diffuser, initial (low) estimates of the port diameters for Secs. 1 and 2 can be made as follows:

$$d_1 \simeq d_3 \left[\frac{Q_{3_p}}{Q_{1_p}}\right]^{1/2} = 2.79 \left(\frac{6.67}{4.01}\right)^{1/2} = 3.60 \text{ in.}$$

$$d_2 \simeq d_3 \left[\frac{Q_{2_p}}{Q_{1_p}}\right]^{1/2} = 2.79 \left(\frac{9.32}{4.01}\right)^{1/2} = 4.25 \text{ in.}$$

These values for d_1, d_2, and d_3 are the trial diameters to be used in the initial iterative solution of the discrete mathematical model.

5. Apply the model of Rawn et al. and determine the required port diameter (see computer code in Appendix I). The computational procedure entails determining the port diameters for Secs. 1, 2, and 3 that will give $Q_p = 20$ cfs and the desired discharge distribution for $E_N = 4.584$ ft. The first set of iterations for the estimated port diameters is summarized in Table 2-6. The desired total discharge of 20 cfs is obtained for $E_1 = 3.665$ ft. The corresponding value of $E_N = 4.510$ is in close agreement with the desired value of 4.584. At the design head of $E_N = 4.584$, the discharge for the system is 20.03 cfs.

    The discharges for Secs. 1, 2, and 3 corresponding to the conditions in the last row of Table 2-6 are 6.81, 9.23, and 3.96 cfs, respectively. These values are all within 2% of the respective desired values of 6.68, 9.32, and 4.01 cfs and would be considered quite acceptable. Nearly perfect flow division can be obtained at this point, however, by making corrections in the diameters based again on Eq. (2-1), where d is proportional to $Q^{1/2}$, as follows

$$d_1 = 3.60 \left(\frac{6.67}{6.81}\right)^{1/2} = 3.57 \text{ in.}$$

$$d_2 = 4.25 \left(\frac{9.32}{9.23}\right)^{1/2} = 4.27 \text{ in.}$$

$$d_3 = 2.79 \left(\frac{4.01}{3.95}\right)^{1/2} = 2.81 \text{ in.}$$

These values for d_1, d_2, and d_3 give Q_{1_p}, Q_{2_p}, and $Q_{3_p} = 6.69$, 9.30, and 4.00 cfs, respectively at $E_1 = 3.654$ ft and $E_N = 4.552$ ft. In practice, standard

TABLE 2-6   *Summary of results of Step 5 (Example Problem 2-2)*

D (ft)	d_1 (in.)	d_2 (in.)	d_3 (in.)	E_1 (ft)	E_N (ft)	Q_p (cfs)
(1)	(2)	(3)	(4)	(5)	(6)	(7)
2	3.60	4.25	2.79	4.000	4.987	20.894
2	3.60	4.25	2.79	3.600	4.488	19.822
2	3.60	4.25	2.79	3.665	4.570	20.000

sizes for the port diameters should be used. Rounding the "newest" d values upward to the nearest sixteenth of an inch, gives d_1, d_2, and $d_3 = 3.5625$, 4.3125, and 2.8125 ($3\frac{9}{16}$, $4\frac{5}{16}$, and $2\frac{13}{16}$) in., respectively. At $E_1 = 3.596$ ft and $E_N = 4.494$ ft, these values yield $Q_{1_p} = 6.61$ cfs, $Q_{2_p} = 9.41$ cfs, and $Q_{3_p} = 3.97$ cfs, which are all within about 1% of the desired values. At the available E_N of 4.584 ft, the system discharge is 20.20 cfs. It would probably be advisable to choose somewhat larger port sizes to allow for increased friction losses with time.

The port discharge and the manifold pipe velocity are given in Table 2-7 for d_1, d_2, and $d_3 = 3\frac{9}{16}$, $4\frac{5}{16}$, and $2\frac{13}{16}$ respectively. Column 7 of Table 2-7 shows the pipe velocity distribution for the average flow rate of 9 cfs. The velocity exceeds 2 fps for about 1/2 of the diffuser length. The variation of port discharge is less than 4% for any of the three sections, as shown in Column 5. The same relative discharge distribution will exist for any total discharge since it is a freshwater diffuser.

EXAMPLE PROBLEM 2-3:   Design the river diffuser described in Example Problem 2-2 for an available head of only 4 ft.

Solution: Steps 1 and 2 will be the same as for Example Problem 2-2. Therefore, the same wastewater discharge distribution and manifold pipe diameter will be used. The basic difference is that, although the head loss along the pipe is the same as before, the variation in the available head at each port, and therefore the port discharge, is much greater for this situation because the value of E_N is significantly smaller. As a result, more iterations will be needed since the initial estimates of the port diameters, obtained by assuming $E = E_N$ everywhere, will be quite rough.

3. Determine the total head at the upstream end of the outflow portion of the diffuser. The head at the Nth port is

$$E_N = H - \frac{8 fL Q_p^2}{\pi^2 g D^5} = 4 - 1.416$$

$$E_N = 2.584 \text{ ft}$$

TABLE 2-7　Port discharge and manifold velocity distributions for final design (Example Problem 2-2) (E_1 = 3.596 ft, E_N = 4.494 ft).

Port no.	Pipe diameter (ft)	Port diameter (in.)	Total Head (ft)	Port discharge (cfs)	Pipe velocity when Q = 20 cfs (fps)	Pipe velocity when Q = 9 cfs (fps)
(1)	(2)	(3)	(4)	(5)	(6)	(7)
1	2		3.596	0.664	0.21	0.10
2	2		3.596	0.664	0.42	0.19
3	2		3.596	0.663	0.63	0.29
4	2		3.597	0.663	0.84	0.38
5	2	3 9/16	3.599	0.662	1.05	0.48
6	2		3.601	0.661	1.26	0.57
7	2		3.604	0.660	1.47	0.66
8	2		3.608	0.659	1.68	0.76
9	2		3.614	0.658	1.89	0.85
10	2		3.621	0.657	2.10	0.95
11	2		3.629	0.960	2.41	1.08
12	2		3.640	0.956	2.71	1.22
13	2		3.655	0.952	3.02	1.36
14	2		3.672	0.948	3.32	1.49
15	2	4 5/16	3.694	0.944	3.62	1.63
16	2		3.719	0.939	3.92	1.76
17	2		3.749	0.935	4.21	1.90
18	2		3.784	0.931	4.51	2.03
19	2		3.823	0.926	4.80	2.16
20	2		3.868	0.923	5.10	2.30
21	2		3.918	0.391	5.22	2.35
22	2		3.971	0.392	5.35	2.41
23	2		4.027	0.393	5.47	2.46
24	2		4.085	0.395	5.60	2.52
25	2	2 13/16	4.146	0.396	5.72	2.58
26	2		4.210	0.398	5.85	2.63
27	2		4.276	0.399	5.98	2.69
28	2		4.346	0.401	6.11	2.75
29	2		4.418	0.403	6.24	2.81
30	2		4.498	0.405	6.37	2.86

4. Select port sizes and spacing so that the desired discharge distribution is obtained at peak flow for the available head. Again, using a port spacing of 10 ft, the port diameter for Sec. 3 can be estimated as follows

$$d_3 = d_N \simeq \frac{4(0.401)}{(0.489)\pi 2g(2.594)} = 3.29 \text{ in.}$$

and, as before,

$$d_1 \simeq d_3(Q_{3_p}/Q_{1_p})^{1/2} = 3.29\left(\frac{6.67}{4.01}\right)^{1/2} = 4.24 \text{ in.}$$

$$d_2 \simeq d_3(Q_{2_p}/Q_{1_p})^{1/2} = 3.29\left(\frac{9.32}{4.01}\right)^{1/2} = 5.02 \text{ in.}$$

5. Apply the model of Rawn et al. and determine the required port diameter (see computer code in Appendix I). The results of the first set of iterations for d_1, d_2, and $d_3 = 4.24$, 5.02, and 3.29 in., respectively are summarized in Table 2-8. The head required to pass the peak design flow for the initial estimates of the port diameters is 0.292 ft greater than the available head. Also, the discharges corresponding to the last line in Table 2-8 for the diffuser sections are $Q_{1_p} = 6.90$, $Q_{2_p} = 9.18$, and $Q_{3_p} = 3.93$, which are about 2 to 3% different from the desired values. In this case, therefore, the diameters should be changed by considering that d is proportional to $q^{1/2} E^{-1/4}$ from Eq. (2-1). This yields the following

$$d_1 = 4.24 \left(\frac{6.61}{6.90}\right)^{1/2} \left(\frac{2.584}{2.876}\right)^{-1/4} = 4.29 \text{ in.}$$

$$d_2 = 5.02 \left(\frac{9.32}{9.18}\right)^{1/2} \left(\frac{2.584}{2.876}\right)^{-1/4} = 5.20 \text{ in.}$$

$$d_3 = 3.29 \left(\frac{4.01}{3.93}\right)^{1/2} \left(\frac{2.584}{2.876}\right)^{-1/4} = 3.41 \text{ in.}$$

Although this approach for correcting the diameters is quite approximate, it does head things in the right direction.

The results of the calculations based on the newest d values are shown in Table 2-9. The discharges for the three 100-ft sections corresponding to the last line in Table 2-9 are $Q_{1_p} = 6.68$ cfs, $Q_{2_p} = 9.31$ cfs, and $Q_{3_p} = 4.02$ cfs, which are in good agreement with the desired values of 6.68, 9.32, and 4.01 cfs, respectively. The required head of 2.678 ft, at the upstream end of the diffuser is still higher than the available head

TABLE 2-8 Summary of results for initial assumed port diameters for Step 5 (Example Problem 2-3).

D (ft)	d_1 (in.)	d_2 (in.)	d_3 (in.)	E_1 (ft)	E_N (ft)	Q_p (cfs)
(1)	(2)	(3)	(4)	(5)	(6)	(7)
2	4.24	5.02	3.29	2.000	2.924	20.165
2	4.24	5.02	3.29	1.900	2.778	19.654
2	4.24	5.02	3.29	1.967	2.876	19.998

TABLE 2-9 Summary of results for corrected port diameters for Step 5 (Example Problem 2-3)

D (ft)	d_1 (in.)	d_2 (in.)	d_3 (in.)	E_1 (ft)	E_N (ft)	Q_p (cfs)
(1)	(2)	(3)	(4)	(5)	(6)	(7)
2	4.29	5.20	3.41	1.800	2.712	20.126
2	4.29	5.20	3.41	1.700	2.561	19.559
2	4.29	5.20	3.41	1.778	2.678	20.002

of 2.584 ft; therefore, each of the ports needs to be enlarged. Adopting the standard-sized pipe diameters $d_1 = 4\frac{3}{8}$ in., $d_2 = 5\frac{5}{16}$ in., and $d_3 = 3\frac{1}{2}$ in. results in the values shown in Table 2-10. The sectional discharges corresponding to the last line in Table 2-10 are $Q_{1_p} = 6.70$, $Q_{2_p} = 9.29$, and $Q_{3_p} = 4.01$, which are quite close to the desired values. The value of $E_N = 2.539$ in Column 6 is also acceptable.

The port discharge, manifold pipe velocity, and head distributions are presented in Table 2-11 for the diffuser design. Note that the port discharge variation for Secs. 2 and 3 is as high as about 9% which is much greater than the variation for the diffuser discussed in Example Problem 2-1. This is to be expected when available head decreases with respect to the total frictional head loss in the diffuser. The port diameter could be varied along the three diffuser sections to "fine tune" the diffuser and decrease the discharge variation as was done for the 200-ft section in Example Problem 2-1.

The pipe velocity distribution for the average wastewater flow rate of 9 cfs is given in Column 7 of Table 2-11.

EXAMPLE PROBLEM 2-4:   Wastewater is to be discharged into a river uniformly from a 200-ft-long cast-iron diffuser pipe. The ports are to be bell mouthed and spaced at 10-ft intervals. The effluent is transported from the treatment plant to the river through a 150-ft-long pipe of the same diameter and material as the diffuser. The entrance to the pipe is abrupt. The maximum allowable head at the treatment plant (with respect to the river surface) is 7.58 ft. Using the continuous method of Camp and Graber [Eqs. (2-18) and (1-31)], design the diffuser pipe system subject to the constraints given previously for peak and average flow rates of 15 and 7 cfs, respectively.

Solution:

1. Determine the diameter of the manifold pipe. The velocity of the middle of the diffuser manifold pipe should be about 2 fps for the average flow rate in order to minimize settling of solid particles in the system. This criterion enables the diameter of the manifold to be determined as follows:

$$D = \sqrt{\frac{\pi Q_{\text{mid}}}{4V_{\text{mid}}}} = \sqrt{\frac{\pi(3.5 \text{ cfs})}{4(2 \text{ fps})}} = 1.49 \approx 1.5 \text{ ft.}$$

   where Q_{mid} = flow rate at middle of diffuser and V_{mid} = specified velocity at middle of diffuser. The rest of the design procedure is based on delivering the peak flow at the available head.

2. Establish a value for the friction factor, f. A value of $f = 0.017$ is obtained from the rough pipe region of Fig. 1-7 for $e/D = 0.0102$ in./18 in. = 0.00057. Since Camp and Graber's method assumes a constant friction factor, $f = 0.017$ will be used throughout the calculations. The Reynolds number dependence of f could also be accounted for here by using some average manifold pipe velocity.

TABLE 2-10 *Summary of results for adopted standard port diameters for Step 5 (Example Problem 2-3).*

D (ft)	d_1 (in.)	d_2 (in.)	d_3 (in.)	E_1 (ft)	E_N (ft)	Q_p (cfs)
(1)	(2)	(3)	(4)	(5)	(6)	(7)
2	4 3/8	5 5/16	3 1/2	1.700	2.631	20.395
2	4 3/8	5 5/16	3 1/2	1.600	2.476	19.750
2	4 3/8	5 5/16	3 1/2	1.641	2.539	20.001

TABLE 2-11 *Port discharge and manifold velocity distributions for final design (Example Problem 2-3).*

Port no.	Pipe diameter (ft)	Port diameter (in.)	Port discharge (cfs)	Pipe velocity (fps)	Total head (ft)	Pipe velocity for Q = 9 cfs (fps)
(1)	(2)	(3)	(4)	(5)	(6)	(7)
1	2		0.676	0.215	1.641	0.097
2	2		0.676	0.430	1.641	0.194
3	2		0.675	0.645	1.641	0.290
4	2		0.674	0.860	1.642	0.387
5	2	4 3/8	0.672	1.074	1.644	0.483
6	2		0.670	1.287	1.646	0.579
7	2		0.668	1.500	1.649	0.675
8	2		0.666	1.712	1.654	0.770
9	2		0.663	1.923	1.659	0.865
10	2		0.660	2.133	1.666	0.960
11	2		0.968	2.441	1.675	1.098
12	2		0.960	2.746	1.687	1.236
13	2		0.951	3.049	1.701	1.372
14	2		0.942	3.349	1.719	1.507
15	2	5 5/16	0.932	3.646	1.741	1.640
16	2		0.923	3.939	1.767	1.773
17	2		0.914	4.230	1.797	1.904
18	2		0.906	4.519	1.832	2.033
19	2		0.899	4.805	1.872	2.162
20	2		0.892	5.089	1.916	2.290
21	2		0.385	5.211	1.967	2.345
22	2		0.388	5.335	2.019	2.401
23	2		0.391	5.459	2.075	2.457
24	2		0.395	5.585	2.132	2.513
25	2	3 1/2	0.399	5.712	2.193	2.570
26	2		0.403	5.840	2.256	2.628
27	2		0.407	5.969	2.322	2.686
28	2		0.411	6.100	2.392	2.745
29	2		0.416	6.233	2.464	2.805
30	2		0.421	6.367	2.539	2.865

3. Compute the piezometric head, h_0, at the upstream end of the 200-ft diffuser pipe. This can be determined for the peak flow rate by applying the energy equation from the treatment plant to the end of the 150-ft pipeline as follows:

$$7.58 = h_0 + \frac{V_0^2}{2g} + 0.5\frac{V_0^2}{2g} + \frac{fL}{D}\frac{V_0^2}{2g}$$

$$h_0 = 7.58 - \left[1.0 + 0.5 + \frac{0.017(150)}{1.5}\right]\frac{V_0^2}{2g}$$

$$h_0 = 7.58 - (3.2)\frac{1}{2g}\left[\frac{15}{\frac{(1.5)^2}{4}}\right]^2 = 7.58 - 3.2\frac{(8.49)^2}{2g}$$

$$h_0 = 4.00 \text{ ft}$$

4. Develop a relationship for change in piezometric head with distance from diffuser inlet. Using the values of $f = 0.017$ and $V_0 = 8.49$ fps determined above, Eq. (2-18) becomes

$$\Delta h_x = h - h_0$$

$$= \frac{(8.49)^2}{2g}\left\{1 - \left(1 - \frac{x}{200}\right)^2 - \frac{\left[\frac{0.017}{1.5}\right](200)}{3}\left[1 - \left(1 - \frac{x}{200}\right)^3\right]\right\}$$

$$= 0.277 - 1.12\zeta^2 + 0.843\zeta^3$$

where $\zeta = 1 - x/200$.

5. The discharge coefficient, C_d, in Eq. (1-31) is assumed constant in Camp and Graber's method. Thus, it can be determined by combining Eqs. (2-3) and (1-31) as follows:

$$C_d = \frac{0.975\left(1 - \frac{V^2}{2gE}\right)^{0.375}}{\left(1 - \frac{V^2}{2gE}\right)^{0.05}} = 0.975\left(1 - \frac{V^2}{2gE}\right)^{-0.125}$$

then using average values of V and E for the diffuser manifold. The total head at the upstream end of the diffuser, E_0, is given by

$$E_0 = h_0 + \frac{V_0^2}{2g} = 4 + \frac{(8.49)^2}{2g} = 5.10 \text{ ft.}$$

The total head at the downstream end of the diffuser, E_{DS}, is equal to the piezometric head at the end since $V = 0$. Therefore, setting $X = L$ in Eq. (2-18),

$$E_{DS} = h_0 + h_L = 4 + \frac{V_0^2}{2g}\left(1 - \frac{FL}{3}\right) = 4 + \frac{(8.49)^2}{2g}\left[1 - \frac{\left[\frac{0.017}{1.5}\right]200}{3}\right]$$

$$= 4.276 \text{ ft}$$

Hence

$$E_{\text{ave}} = \frac{E_0 + E_{DS}}{2} = 4.70 \text{ ft}$$

The velocity at the middle of the pipe, $8.49/2 = 4.25$ fps, is used for the average velocity. Consequently, the discharge coefficient is calculated as follows

$$C_d = 0.975\left[1 - \frac{(4.25)^2}{2g(4.70)}\right]^{-0.125} = 0.983$$

6. Determine the port size by assuming an equal discharge per port of

$$q = Q_{\text{peak}}/(\text{No. of ports}) = \frac{15 \text{ cfs}}{20} = 0.75 \text{ cfs}$$

and from Eq. (1-31)

$$d = \left[\frac{4g}{\pi C_d}\frac{1}{\sqrt{2g\,\Delta h_{\text{ave}}}}\right]^{1/2} = \left[\frac{4(0.75)}{\pi(0.983)}\frac{1}{[64.4(4.14)]^{1/2}}\right]^{1/2}$$

$$d = 0.244 \text{ ft} = 2.94 \text{ in.}$$

or, rounding to standard units, $d = 3$ in.

7. Establish the discharge distribution of the proposed design mathematically using the following procedure for each port:

   (a) Determine x of port

   (b) Calculate $\Delta h_x = 0.277 - 1.12\zeta^2 + 0.843\zeta^3$

      where $\zeta = (1 - x/200)$

   (c) Calculate $h = h_0 + \Delta h_x = 4 + \Delta h_x$

   (d) Calculate $q = C_d (\pi d^2/4) \sqrt{2g\,\Delta h} = 0.387\,\Delta h^{1/2}$

Table 2-12 summarizes these calculations for the proposed design. Note that each port is in the middle of a 10-ft pipe section. The total discharge for peak conditions is 15.7 cfs. This indicates that the design is on the safe side relative to overtopping the containment at the treatment plant. A larger factor of safety could be realized, however, by making the ports larger than 3 in. The discharge distribution is about ± 2%, which is quite acceptable.

TABLE 2-12 *Summary of calculation procedure and discharge distribution at peak flow for Example Problem 2-4.*

Port no.	x (ft)	$\zeta = 1 - x/L$	Δh_x (ft)	Δh (ft)	q (cfs)
(1)	(2)	(3)	(4)	(5)	(6)
20	5	0.975	−0.006	3.994	0.773
19	15	0.925	−0.014	3.986	0.773
18	25	0.875	−0.016	3.984	0.773
17	35	0.825	−0.012	3.988	0.773
16	45	0.775	−0.003	3.997	0.774
15	55	0.725	+0.010	4.010	0.775
14	65	0.675	+0.026	4.026	0.777
13	75	0.625	+0.045	4.045	0.778
12	85	0.575	+0.068	4.068	0.781
11	95	0.525	+0.090	4.090	0.783
10	105	0.475	+0.115	4.115	0.785
9	115	0.425	+0.139	4.139	0.787
8	125	0.375	+0.164	4.164	0.790
7	135	0.325	+0.188	4.188	0.792
6	145	0.275	+0.210	4.210	0.794
5	155	0.225	+0.230	4.230	0.796
4	165	0.175	+0.247	4.247	0.798
3	175	0.175	+0.261	4.261	0.799
2	185	0.075	+0.271	4.271	0.800
1	195	0.025	+0.276	4.276	0.800

$$\Sigma q = 15.699 \text{ cfs}$$

REFERENCES

CAMP, T. R., and GRABER, S. D. "Dispersion Conduits," *Journal of the Sanitary Engineering Division, ASCE,* **94,** 169 (1968).

FISCHER, H. B., LIST, E. J., KOH, C. Y., IMBERGER, J., and BROOKS, N. H., *Mixing in Inland and Coastal Waters,* Academic Press, Inc., New York, New York 1979.

FRENCH, J. A., "Internal Hydraulics of Multiport Diffuser," *JWPCF,* **44,** 782 (1972).

JAIN, S. C., SAYRE, W. W., AKYEAMPONG, Y. A., McDOUGALL, D., and KENNEDY, J. F., "Model Studies and Design of Thermal Outfall Structures Quad Cities Nuclear Plant," *IIHR Report No. 135,* University of Iowa, Iowa City, Iowa, September 1971.

McKNOWN, J. S., and HSU, E. Y., "Application of Conformal Mapping to Divided Flow," *Proceedings of the First Midwestern Conference on Fluid Mechanics,* J. W. Edwards, Ann Arbor, Michigan, 1951.

McKNOWN, J. S., "Mechanics of Manifold Flow," *Transactions, ASCE,* **119,** 1111 (1954).

RAWN, A. M., BOWERMAN, F. R., and BROOKS, N. H., "Diffusers for Disposal of Sewage in Sea Water," *Transactions, ASCE,* **126,** Part III, 344 (1961).

3

FLOW IN OPEN CHANNELS

Open channel flow occurs when water is transported through flumes, canals, ditches, etc., and is very common in environmental engineering applications. The upper boundary of the open channel flow cross section is the free water surface. At this boundary an air–water interface exists, and the pressure is atmospheric. Therefore, for open channel flow, the hydraulic grade line coincides with the water surface. Hence, open channel flow is significantly different from full-bore pipe flow. In pipe flow, pressure causes the water to move through the pipe, and the flow is basically independent of the location of the pipe. However, in open channel flow, water movement is due to gravitational forces that drive water from higher elevations to lower elevations. The variable cross-sectional area characteristic of open channel flow also adds an additional unknown to the basic flow equations.

The flow in an open channel may be laminar or turbulent, uniform or varied (nonuniform), steady or unsteady. Most design problems encountered in practice are steady, uniform, turbulent flow or steady, varied, turbulent flow. Uniform flow in open channels occurs when the cross-sectional areas of flow and depth remain constant over a specified reach of channel. Varied flow occurs when the depth of flow changes from section to section in a channel (this is not a time variation). Normally, uniform flow will occur only in a long channel after a transition from varied flow conditions has occurred (varied flow will occur in bends, at changes in channel slope, at changes in cross-sectional area, or at changes in channel roughness). Still, the fundamental uniform flow equations are very important because many varied flow problems are solved by discrete application of uniform flow theory.

3.1 UNIFORM FLOW

A flow schematic for a short reach of an open channel with uniform flow is presented in Fig. 3-1 (y_o represents the depth of flow and is referred to as the normal depth). Because there is no variation in velocity or depth of flow over the reach, it is necessary

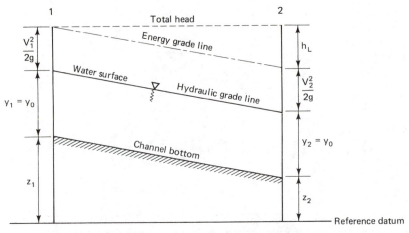

FIGURE 3-1 *Uniform open channel flow.*

that the decrease in potential energy resulting from the change in bed elevation exactly accounts for the energy loss due to friction along the channel walls. Hence, the head loss due to friction, h_L, between point 1 and point 2 on Fig. 3-1, is the difference in the elevation of the energy grade line between these points.

The slope of the bed and the slope of the water surface (both are the same since they are parallel) are equal to the energy gradient for uniform flow, where the energy gradient is defined as

$$S_E = \frac{h_L}{L} \tag{3-1}$$

where S_E = energy gradient, ft/ft

   h_L = head loss over specified length of channel, ft

   L = length measured along the channel (not necessarily the horizontal), ft

In 1775 Chézy proposed that the average velocity for the special case of *uniform open channel flow* was given by

$$V = C_c \sqrt{RS_E} \tag{3-2}$$

where V = average velocity, ft/s

   R = hydraulic radius, which is obtained by dividing the cross-sectional area of flow by the wetted perimeter, ft

   C_c = Chézy discharge coefficient, ft$^{1/2}$/s

The Chézy discharge coefficient is obtained from the relationship

$$C_c = \sqrt{\frac{8g}{f}} \tag{3-3}$$

where f = Darcy–Weisbach friction factor

g = gravity constant, ft/s^2

In 1889 Manning proposed the equation

$$V = \frac{1.49}{n} R^{2/3} S_E^{1/2} \tag{3-4}$$

or

$$Q = AV = A\frac{1.49}{n} R^{2/3} S_E^{1/2} \tag{3-5}$$

where Q = volumetric flow rate, ft^3/s

A = cross-sectional area of flow, ft^2

n = Manning roughness coefficient, s/ft$^{1/3}$

The Manning roughness coefficient is related to the Darcy–Weisbach friction factor by the equation

$$n = 0.093 f^{1/2} R^{1/6} \tag{3-6}$$

The Manning equation is the equation most commonly applied to uniform flow problems, where the bed slope equals the energy gradient. However, the value of the Manning roughness coefficient is critical if accurate answers are to be obtained with this equation. Hence, every effort should be made to ensure that an appropriate value for n is selected. Typical values for n are presented in Table 1-4.

Circular Conduits Flowing Partly Full

When a circular conduit is flowing partially full, the principles of open channel flow apply because the surface of the water is normally exposed to atmospheric pressure. Consider the flow schematic presented in Fig. 3-2. The angle Θ, the cross-sectional area of flow, and the wetted perimeter are given by the following relationships:

$$\cos\frac{\Theta}{2} = 1 - \frac{2y_0}{D} \tag{3-7}$$

$$A = \frac{D^2}{4}\left[\frac{\pi\Theta}{360} - \frac{\sin\Theta}{2}\right] \tag{3-8}$$

$$WP = \frac{\Theta}{360}(2\pi r) \tag{3-9}$$

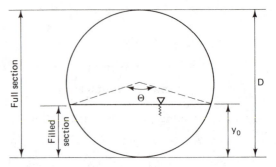

FIGURE 3-2 *Flow in a partially filled circular conduit.*

where D = diameter of circular section, ft

r = radius of circular section, ft

WP = wetted perimeter, ft

If Manning's roughness coefficient, n, is assumed to remain constant with depth, Manning's equation and the continuity equation may be used to compute the average velocity and discharge for various depths of flow. Values obtained from such computations may be compared to the values obtained when the conduit is flowing full, and a graphical representation such as that shown in Fig. 3-3 may be developed.

In reality, n does not remain constant but varies appreciably with the depth of flow. The n distribution reported by Camp (1946) is presented in Fig. 3-4. Although it is common practice to assume that n remains constant when solving partial flow problems, the reader should be aware that n does vary with depth of flow and that computational accuracy is enhanced when this variation is considered.

Dake (1972) has shown that the best performance of a circular section occurs when

$$4\Theta - 6\Theta(\cos \Theta) + \sin 2\Theta = 0 \qquad \textbf{(3-10)}$$

The solution of Eq. (3-10) gives $\Theta \simeq 154°$. For this angle, Eq. (3-7) may be solved for y_0 to give

$$y_0 \simeq 1.9r \qquad \textbf{(3-11)}$$

This means that a circular section discharges most efficiently when it is flowing about 95% full. In other words, a circular section passes its maximum flow at $y_0 \simeq 1.9r$. According to Dake (1972), the type of flow changes from free surface flow to pressure flow when the depth of flow exceeds $1.9r$. This causes the carrying capacity of the pipe to decrease and water to accumulate to provide the head necessary for pressure flow. Actually, 95% open channel pipe flow would rarely be a stable condition. In sewer pipes, for instance, the flow rate surges, or changes, all the time and the flow oscillates continuously between pressure flow and open channel flow.

EXAMPLE PROBLEM 3-1: Compute the average velocity and the discharge through an 18-in.-diameter pipe placed on a slope of 0.003 ft/ft if the n value for clean uncoated cast-iron pipe is assumed to be 0.013 (see Table 1-4) and the depth of flow in the pipe is 10 in.

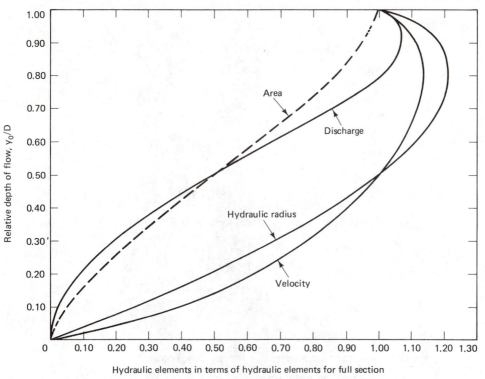

Hydraulic elements in terms of hydraulic elements for full section

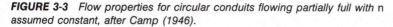

$$\frac{V}{V_{full}}, \quad \frac{Q}{Q_{full}}, \quad \frac{A}{A_{full}}, \text{ and } \frac{R}{R_{full}}$$

FIGURE 3-3 *Flow properties for circular conduits flowing partially full with* n *assumed constant, after Camp (1946).*

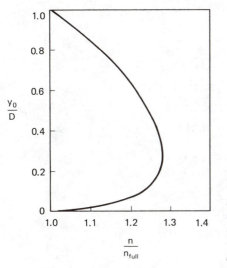

FIGURE 3-4 *Normalized distribution of Manning's* n *nersus relative depth in a circular section, from Camp, T. R., "Design of Sewers to Facilitate Flow," Sewage Works Journal, 18, 1, 1946. Used with permission of Water Pollution Control Federation.*

Solution:

1. Compute the cross-sectional area of flow.

    (a) Apply Eq. (3-7) and determine Θ.

$$\cos \frac{\Theta}{2} = 1 - \frac{2y_0}{D}$$

$$= 1 - \frac{(2)(10)}{18}$$

or

$$\Theta = 160°$$

    (b) Calculate the cross-sectional area from Eq. (3-8).

$$A = \frac{D^2}{4}\left[\frac{\pi\Theta}{360} - \frac{\sin\Theta}{2}\right]$$

$$= \frac{(18/12)^2}{4}\left[\frac{\pi(160)}{360} - \frac{\sin(160)}{2}\right]$$

$$= 0.72 \text{ ft}^2$$

2. Determine the hydraulic radius for the flow section.

    (a) Compute the wetted perimeter from Eq. (3-9).

$$WP = \frac{\Theta}{360}(\pi D)$$

$$= \left[\frac{160}{360}\right](\pi)(18/12)$$

$$= 2.09 \text{ ft}$$

    (b) Calculate the hydraulic radius.

$$R = \frac{A}{WP}$$

$$= \frac{0.72}{2.09}$$

$$= 0.34 \text{ ft}$$

3. Estimate the average velocity from Eq. (3-4).

$$V = \frac{1.49}{n}R^{2/3}S_E^{1/2}$$

$$= \frac{1.49}{0.013}(0.34)^{2/3}(0.003)^{1/2}$$

$$= 3.06 \text{ ft/s}$$

4. Determine the discharge by applying the continuity equation.

$$Q = AV$$
$$= (0.72)(3.06)$$
$$= \mathbf{2.2 \ ft^3/s}$$

If Q had been given and it were required to compute the average velocity and depth of flow, a trial and error solution would be necessary. In this case a depth of flow could be assumed and the velocity and discharge associated with this depth computed. The problem is solved when the assumed depth gives a computed discharge equal to the given discharge.

3.2 VARIED FLOW

Probably the most commonly encountered type of flow in open channels is steady (constant rate of discharge) and varied (changing in depth and velocity). Since the depth of flow is continually changing, the slope of the water surface will not be parallel to the channel bottom.

For the purpose of analytical treatment, varied flow is classified as (a) gradually varied flow (GVF) and (b) rapidly varied flow (RVF). The primary difference between these two types of varied flow is that in *gradually varied flow* the flow conditions change slowly over a long reach of channel, whereas in *rapidly varied flow* the flow conditions change abruptly and normally affect only a short reach of channel. It should be remembered that GVF is dominated by friction, whereas RVF is dominated by inertial forces and friction is usually neglected.

Specific Energy

For an open channel of small slope, the total energy per unit weight of fluid at any section, for any arbitrary streamline, is the total head (recall total head is the sum of the velocity head, the pressure head, and the elevation head). When the channel bottom is chosen as the datum, the total head or energy is called the *specific energy*, E. Specific energy may be expressed mathematically as

$$E = \frac{P}{\gamma} + Z + \frac{V^2}{2g} \qquad (3\text{-}12)$$

where Z is the elevation from the bed. However, for any streamline in the section $(P/\gamma) + Z = y$. Hence, Eq. (3-12) may be written as

$$E = y + \frac{V^2}{2g} \qquad (3\text{-}13)$$

For the case of uniform flow the energy line is parallel to the channel bottom, and the specific energy is constant. However, for varied flow the energy line is not parallel to the channel bottom, although it always slopes downward in the direction of flow, and depending upon the particular flow conditions and channel geometry, the specific energy may increase or decrease.

Critical Depth

Equation (3-13) may be rewritten so that specific energy is expressed in terms of discharge.

$$E = y + \frac{Q^2}{2gA^2} \tag{3-14}$$

Consider a situation where the discharge, Q, is held constant but the depth of flow is varied by some means, e.g., changing the channel cross-section or bed elevation. A plot of depth versus specific energy for this case is shown in Fig. 3-5. Specific energy is the sum of the kinetic energy (due to velocity) and the potential energy (due to depth of flow). These two energy components are summed on the depth versus specific energy diagram to give a trace on which specific energy decreases with decreasing depth until point C on the curve is reached. After point C, specific energy increases as the depth decreases. The depth and velocity associated with point C are termed the *critical depth* and the *critical velocity*. For a channel of a given shape, the

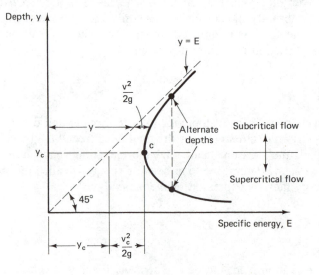

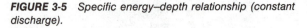

Note: The y = E trace will be inclined
at 45° when the same scale is used
for the y-axis and the x-axis.

FIGURE 3-5 *Specific energy–depth relationship (constant discharge).*

specific energy associated with the critical depth is the *minimum* specific energy able to maintain the specified discharge. If the specific energy is greater than the minimum value, two different depths of flow are possible. One of the depths is greater than the critical depth, whereas the other depth is less than the critical depth. These are referred to as *alternate depths* or *alternate stages*. Flow associated with the alternate depth greater than critical depth is called *subcritical flow,* whereas the flow associated with the alternate depth less than critical depth is called *supercritical flow*.

Since critical depth occurs when specific energy is at a minimum, the conditions for critical depth may be obtained by setting the derivative of specific energy, with respect to depth, equal to zero. By considering Eq. (3-14), it can be shown that for critical depth to occur, the following relationship must be satisfied irrespective of channel shape (Webber, 1965):

$$\frac{Q^2 W}{gA^3} = 1 \qquad (3\text{-}15)$$

where W = width of the water surface, ft

Since W and A in Eq. (3-15) have values associated with critical depth, Eq. (3-15) is more appropriately represented as

$$\frac{Q^2 W_c}{gA_c^3} = 1 \qquad (3\text{-}16)$$

Consider the case of a *rectangular channel* where

$$A = Wy \qquad (3\text{-}17)$$

Substituting for A in Eq. (3-15) from Eq. (3-17) and solving for y_c gives

$$y_c = \left[\frac{Q^2}{gW^2} \right]^{1/3} \qquad \textit{(rectangular channel)} \qquad (3\text{-}18)$$

If S_c is defined as the *critical slope,* i.e., the slope of the channel bottom that will produce critical uniform flow, a channel slope less than S_c is called a *mild slope* and will produce subcritical uniform flow for the same discharge and roughness. Conversely, a slope greater than S_c is called a *steep slope* and will produce supercritical uniform flow. It is important to note, however, that varied subcritical, critical, or supercritical flow can occur on any slope. Therefore, slope classification depends on discharge and roughness as well as on bed slope for a given geometry.

Channel Transitions

It is convenient to classify open channel transitions on the basis of flow transformation:

1.   Subcritical flow transformed to another subcritical flow level.

2. Subcritical flow transformed to supercritical flow.

3. Supercritical flow transformed to subcritical flow.

4. Supercritical flow transformed to another supercritical flow level.

Cases 2 and 3 are of considerable importance, because when flow changes from subcritical to supercritical or from supercritical to subcritical, the depth of flow must pass through critical depth.

A channel transition that transforms the flow from subcritical to supercritical is illustrated in Fig. 3-6. In this figure the channel transition is located at the point where the slope of the channel bottom is abruptly increased from a value less than critical slope to a value greater than critical slope. It has also been found that critical depth will occur in a channel with subcritical flow upstream from a free overfall (see Fig. 3-7). According to Daugherty and Franzini (1977), critical depth will occur some distance upstream from the free overfall. This distance will generally have a value between $3y_c$ and $10y_c$. The depth of flow at the brink of the free overfall has a value approximately equal to $0.72\ y_c$.

In any situation where the flow is transformed from subcritical to supercritical, a *control section* forms. The control section is located at the point where critical depth occurs and is so called because the depth of flow at the control section controls the depth of flow *upstream* and *downstream* from this section.

A channel transition that transforms the flow from supercritical to subcritical is illustrated in Fig. 3-8. In this case the slope of the channel bottom is abruptly reduced

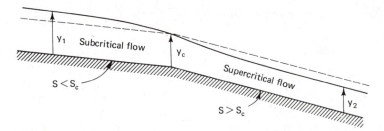

FIGURE 3-6 *Channel transition where flow is transformed from subcritical to supercritical, from* Fluid Mechanics with Engineering Applications, seventh edition, *by Daugherty and Franzini, 1977. Used with permission of McGraw-Hill Book Company.*

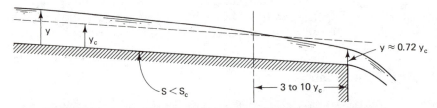

FIGURE 3-7 *Free overfall channel transition where flow is transformed from subcritical to supercritical, from* Fluid Mechanics with Engineering Applications, seventh edition, *by Daugherty and Franzini, 1977. Used with permission of McGraw-Hill Book Company.*

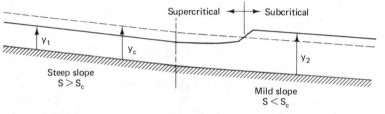

FIGURE 3-8   *Transition from supercritical to subcritical flow.*

from a value greater than the critical slope to a value less than the critical slope. The flow upstream from the point where the slope of the channel bottom changes is supercritical, whereas a short distance downstream from this point the flow becomes subcritical. Within this short distance, the water surface elevation rises rapidly, passing through critical depth and stabilizing at some subcritical depth. This phenomenon is known as a *hydraulic jump*.

To understand why the hydraulic jump occurs, consider the depth versus specific energy diagram presented in Fig. 3-5. Some unique depth (less than critical depth) is associated with critical flow. When the flow passes over the transition from a steep slope to a mild slope (Fig. 3-8), frictional resistance retards the high velocity of flow, resulting in a loss of specific energy and an increase in the depth of flow. Such a response can raise the depth of flow to the critical depth, but, as seen in Fig. 3-5, for the depth of flow to increase beyond critical depth, energy would have to be added from an external source. However, this does not occur. Instead, before the critical specific energy is reached, the hydraulic jump is initiated. During the jump, highly turbulent conditions exist that result in large amounts of air being entrained into the water. This reduces the density of the water slightly, causing a rapid rise in the water surface elevation to the appropriate subcritical flow depth. A considerable amount of energy is dissipated when a hydraulic jump occurs. Because of this, the specific energy associated with the subcritical flow depth will be significantly less than the specific energy associated with the initial depth at supercritical flow.

Consider the hydraulic jump schematic presented in Fig. 3-9, in which the subscripts 1 and 2 refer to conditions before and after the jump. From momentum

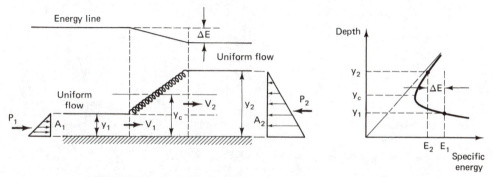

FIGURE 3-9   *Schematic of hydraulic jump, after Webber (1965).*

principles, it can be shown that (Webber, 1965)

$$A_2 \bar{y}_2 - A_1 \bar{y}_1 = \frac{Q^2}{gA_1A_2}(A_2 - A_1) \qquad \text{(3-19)}$$

where $\bar{y}_1$ = distance from the water surface to the centroid of area A_1, ft (not to the centroid of the pressure prism)

$\bar{y}_2$ = distance from the water surface to the centroid of area A_2, ft

The energy dissipated in the hydraulic jump may be obtained from the relationship

$$E_1 - E_2 = \Delta E = (y_1 - y_2) + \left[\frac{V_1^2 - V_2^2}{2g}\right] \qquad \text{(3-20)}$$

or

$$\Delta E = (y_1 - y_2) + \frac{Q^2}{2gA_1^2A_2^2}(A_2^2 - A_1^2) \qquad \text{(3-21)}$$

The depths y_1 and y_2 are referred to as *conjugate depths*. For a *rectangular channel* it is seen that

$$A_1 = Wy_1$$

$$\bar{y}_1 = \frac{y_1}{2}$$

$$A_2 = Wy_2$$

$$\bar{y}_2 = \frac{y_2}{2}$$

Substituting from the above expressions into Eq. (3-19) and simplifying, it is possible to derive equations for the depths of flow

$$y_1 = \left[\frac{y_2^2}{4} + \frac{2q^2}{gy_2}\right]^{1/2} - \frac{y_2}{2} \qquad \text{(3-22)}$$

or

$$y_2 = \left[\frac{y_1^2}{4} + \frac{2q^2}{gy_1}\right]^{1/2} - \frac{y_1}{2} \qquad \text{(3-23)}$$

where q = discharge per unit width of channel, i.e., Q/W, ft^2/s

The ratio of the conjugate depths may be expressed as

$$\frac{y_2}{y_1} = \left[\frac{1}{4} + 2F_1{}^2\right]^{1/2} - \frac{1}{2} \tag{3-24}$$

where F_1 is the upstream Froude number and is given by

$$F_1 = \frac{V_1}{\sqrt{gy_1}} \tag{3-25}$$

If the jump is to occur, F_1 must be greater than one, i.e., the upstream flow must be supercritical.

The energy dissipated in a hydraulic jump in a rectangular channel may also be computed from the equation

$$E_1 - E_2 = \frac{(y_2 - y_1)^3}{4y_1y_2} \tag{3-26}$$

EXAMPLE PROBLEM 3-2:   Is a hydraulic jump possible in a rectangular channel 8 ft wide discharging 400 cfs of water with a depth of flow of 2.0 ft? If the jump occurs, what will be the depth of flow after the jump and what will be the horsepower loss through the jump?

Solution:

1. Compute the upstream Froude number and check to see if it is greater than one.

   (a) Calculate the average velocity of flow in the channel.

   $$V_1 = \frac{Q}{A_1}$$

   $$= \frac{400}{8 \times 2}$$

   $$V_1 = \textbf{25 ft/s}$$

   (b) Compute the upstream Froude number from Eq. (3-25).

   $$F_1 = \frac{V_1}{\sqrt{gy_1}}$$

   $$= \frac{25}{\sqrt{32.2 \times 2}}$$

   $$F_1 = \textbf{3.1}$$

   Since F_1 is greater than one, the flow is supercritical and a hydraulic jump is possible.

2. Determine the depth of flow after the jump by applying Eq. (3-23).

 (a) Estimate q, the discharge per unit width of channel.

$$q = \frac{Q}{W}$$

$$= \frac{400}{8}$$

$$q = 50 \text{ ft}^2/\text{s}$$

 (b) Compute the depth after the jump.

$$y_2 = \left[\frac{y_1^2}{4} + \frac{2q^2}{gy_1}\right]^{1/2} - \frac{y_1}{2}$$

$$= \left[\frac{(2)^2}{4} + \frac{2(50)^2}{(32.2)(4)}\right]^{1/2} - \frac{2}{2}$$

$$= \textbf{5.3 ft}$$

3. Evaluate the horsepower lost through the jump.

 (a) Calculate the average velocity of flow in the channel after the jump.

$$V_2 = \frac{Q}{A_2}$$

$$= \frac{400}{8 \times 5.3}$$

$$V_2 = \textbf{9.43 ft/s}$$

 (b) Compute the total energy loss.

$$\Delta h = \Delta E = E_1 - E_2$$

$$E_1 = 2 + \frac{(25)^2}{(2)(32.2)} = 11.70 \text{ ft}$$

$$E_2 = 5.3 + \frac{(9.43)^2}{(2)(32.2)} = 6.68 \text{ ft}$$

$$\Delta h = 11.70 - 6.68 = \textbf{5.02 ft}$$

 (c) Horsepower loss is given by the expression

$$(hp)_{\text{loss}} = \frac{(\Delta h)(\gamma_w)Q}{550}$$

Hence

$$(hp)_{\text{loss}} = \frac{(5.02)(62.4)(400)}{550}$$

$$= 227.8$$

Gradually Varied Flow

Up to this point in the discussion only steady, uniform flow and rapidly varied flow where changes in cross section occur in a relatively short distance have been considered. In gradually varied flow, the changes in the cross-sectional area of flow occur slowly over a long reach of channel, and backwater curves and drawdown curves (also called backwater curves) are used to describe the water surface profile. Whereas boundary friction was neglected in the analysis of rapidly varied flow, it plays a dominant role in gradually varied flow.

An expression for the depth variation in gradually varied flow has the form

$$\frac{dy}{dx} = \left[\frac{S - S_E}{1 - \dfrac{Q^2 W}{g A^3}} \right] \tag{3-27}$$

When dy/dx has a positive value, it means that depth is increasing downstream. On the other hand, when dy/dx has a negative value, it means that depth is decreasing downstream.

Computation of Backwater Curves

A number of different methods have been used to compute backwater curves, e.g., direct integration, graphical integration, and direct-step method. The presentation here will be limited to the direct-step method. In this method it is assumed that for gradually varied flow it is possible to select a reach of channel short enough so that the slope of the total energy line for this distance is equal to the average of the slope of the total energy lines corresponding to uniform flow at the beginning and end of the reach $[(S_E)_{\text{Ave}} = (S_{E1} + S_{E2})/2]$. Writing Bernoulli's equation between points 1 and 2 shown on Fig. 3-10 gives

$$\frac{V_1^2}{2g} + y_1 + \Delta Z = \frac{V_2^2}{2g} + y_2 + h_L \tag{3-28}$$

where ΔZ = elevation change in the channel bottom between points 1 and 2. ΔZ is given by $S \Delta X$.

   h_L = change in elevation of the energy line between points 1 and 2. h_L is given by $(S_E)_{\text{Ave}} \Delta X$.

$(S_E)_{\text{Ave}}$ = slope of the energy line between points 1 and 2.

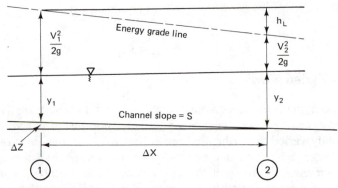

FIGURE 3-10   *Gradually varied flow over a short reach of channel.*

Since

$$E_1 = \frac{V_1^2}{2g} + y_1$$

$$E_2 = \frac{V_2^2}{2g} + y_2$$

Eq. (3-28) may be expressed as

$$E_1 + S\Delta X = E_2 + (S_E)_{\text{Ave}}\Delta X \tag{3-29}$$

Solving Eq. (3-29) for ΔX produces an equation of the form

$$\Delta X = \frac{E_1 - E_2}{(S_E)_{\text{Ave}} - S} \tag{3-30}$$

The shape of a backwater curve may be estimated by employing the following prodedure:

1. Establish the initial conditions of depth, channel geometry, and discharge for point 1.

2. Assume a depth for point 2 and compute the cross-sectional area of flow, the wetted perimeter, the hydraulic radius, the average velocity of flow, and the velocity head.

3. Compute the average slope of the energy line between points 1 and 2.

$$(S_E)_{\text{Ave}} = \frac{n^2(V_1 + V_2)^2}{8.83(R_{\text{Ave}})^{4/3}} \tag{3-31}$$

where $R_{\text{Ave}} = 1/2\ (R_1 + R_2)$ \hfill (3-32)

4. Compute the distance between points 1 and 2.

$$\Delta X = \frac{\left[y_1 + \dfrac{V_1^2}{2g}\right] - \left[y_2 + \dfrac{V_2^2}{2g}\right]}{(S_E)_{\text{Ave}} - S} \tag{3-33}$$

5. Repeat the process for each reach, adding the ΔX values to obtain the total distance required $\Sigma\Delta X$, until the desired depth or desired distance is reached.

EXAMPLE PROBLEM 3-3:   A rectangular channel 2.5 ft wide with a discharge of 12 cfs terminates in a free overfall. If the slope of the channel is 0.001, determine (a) the distance from the free overfall to the point where critical depth occurs and (b) the distance from the free overfall to the point where normal depth occurs. Assume n is constant and has a value of 0.013

Solution:

1. Determine the normal depth, the critical depth, and the depth at the brink.

    (a) The normal depth is obtained from the Manning equation.

    $$Q = A\left[\frac{1.486}{n}\right]R^{2/3}S^{1/2}$$

    which for rectangular channels may be written as

    $$Q = Wy_0\left[\frac{1.486}{n}\right]\left[\frac{Wy_0}{2y + W}\right]^{2/3}S^{1/2}$$

    A trial and error solution shows $y_0 \simeq$ **1.7 ft**

    (b) The critical depth in a rectangular channel is given by Eq. (3-18).

    $$y_c = \left[\frac{Q^2}{gW^2}\right]^{1/3}$$

    $$= \left[\frac{(12)^2}{(32.2)(2.5)^2}\right]^{1/3}$$

    $$y_c = \textbf{0.895 ft}$$

    (c) The depth at the brink is taken as $0.72\ y_c$, hence

    $$y_B = (0.72)(0.89)$$

    $$= \textbf{0.644 ft}$$

2. The problem is solved by beginning with critical depth and, by applying the direct-step method, moving upstream until normal depth is reached. Calculations for a step size of 0.1 ft are summarized in the table on page 96.

y_1 (ft)	y_2 (ft)	V_1 (ft/s)	V_2 (ft/s)	R_{Ave} (ft)	S_{Ave}	ΔX (ft)	$X = \Sigma \Delta X$ (ft)
0.995	0.895	4.822	5.360	0.537	0.00453	4.21	4.21
1.095	0.995	4.381	4.822	0.568	0.00343	15.24	19.45
1.195	1.095	4.015	4.381	0.597	0.00268	31.11	50.56
1.295	1.195	3.705	4.015	0.623	0.00214	55.21	105.77
1.395	1.295	3.439	3.705	0.647	0.00174	95.20	200.97
1.495	1.395	3.209	3.439	0.670	0.00144	172.79	373.76
1.595	1.495	3.008	3.209	0.690	0.00121	383.15	756.91
1.695	1.595	2.831	3.008	0.710	0.00103	2899.68	3656.56

Total distance from point of critical depth to normal depth = $\Sigma \Delta X$ = 3656.6 ft.

In practice the calculations are normally terminated when the depth is within 5% of normal depth. The problem with carrying the calculations out until normal depth is reached is that as normal depth is approached, the calculated distances from Eq. 3-33 become quite large as the denominator approaches zero. This can be seen by noting the difference between the ΔX values in the last two rows in the above table. In this case a more realistic answer is obtained if the calculations are terminated with ΔX = 383.15 and the total distance from the point of critical depth to normal depth is assumed to be 756.91 ft rather than 3656.6 ft.

3. The distance from critical depth to the brink cannot be computed using the direct-step method, because this method is based on gradually varied flow, whereas the flow in the region of interest is rapidly varied. Hence, the distance can only be approximated. Usually, a value of $4y_c$ is assumed.

$$\begin{bmatrix} \text{Distance from} \\ \text{critical depth} \\ \text{to brink} \end{bmatrix} \simeq 4y_c$$

$$\simeq 4(0.895)$$

$$\simeq 3.58 \text{ ft}$$

See Appendix I for the Fortran computer code for this problem.

Lateral Spillway Channels

As stated by Camp (1940), " . . . the term *lateral spillway channel* is used to designate an open channel which receives inflow throughout its length, laterally from one or both sides, and discharges the accumulated flow at a point in the channel, usually one end." Examples of lateral spillway channels encountered in treatment plant design are wash water troughs and gullets for water filters and effluent channels (launders) for sedimentation basins.

Consider the flow schematic presented in Fig. 3-11. In this figure Δy represents the difference in the water depth (measured from the channel bottom) between sections 1 and 2. The $\Delta y'$ term represents the difference in the water surface elevation

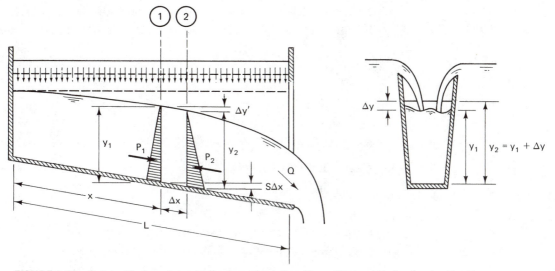

FIGURE 3-11   *Schematic for a lateral inflow problem, from* Open-Channel Hydraulics *by Ven T. Chow, 1959. Used with permission of McGraw-Hill Book Company.*

(measured from the horizontal) between sections 1 and 2 (see Fig. 3-12). Hence,

$$\Delta y' = (y_1 + S\Delta X) - y_2 \tag{3-34}$$

Since

$$\Delta y = y_2 - y_1 \tag{3-35}$$

substituting for $y_1 - y_2$ in Eq. (3-35) gives

$$\Delta y_1' = -\Delta y + S\Delta X \tag{3-36}$$

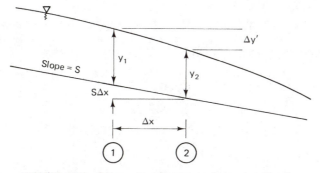

FIGURE 3-12   *Schematic of water surface elevation for lateral inflow.*

Chow (1959) has shown that for a uniform velocity distribution, the *drop* in the water surface elevation between sections 1 and 2 (see Fig. 3-11) may be expressed as

$$\Delta y' = \frac{Q_1(V_1 + V_2)}{g(Q_1 + Q_2)}\left[\Delta V + \frac{V_2}{Q_1}\Delta Q\right] + (S_E)_{Ave}\Delta X \qquad (3\text{-}37)$$

where $\Delta y'$ = drop in water surface elevation between sections 1 and 2, ft

Q_1 = discharge at section 1, ft^3/s

Q_2 = discharge at section 2, ft^3/s

V_1 = velocity at section 1, ft/s

V_2 = velocity at section 2, ft/s

ΔV = change in velocity between section 1 and section 2, i.e., $V_2 - V_1$

ΔQ = change in discharge between section 1 and section 2, i.e., $Q_2 - Q_1$

$(S_E)_{Ave}$ = slope of the energy line as given by Eq. (3-31), ft/ft

ΔX = distance between sections 1 and 2, ft

g = gravity constant, ft/s^2

The first term on the right-hand side of Eq. (3-37) represents the energy loss due to impact as the water falls into the channel. The second term on the right-hand side of Eq. (3-37) represents the energy loss due to friction.

A numerical integration procedure employing Eq. (3-36) and (3-37) may be used to compute the location of the water surface for lateral inflow problems. However, such computations must start at a point of *known depth*. Two types of lateral spillway channels are normally encountered at water and wastewater treatment plants. These are described as follows:

1. The depth of flow at the lower end of the channel is fixed at upper stage by a downstream control point.

2. Critical flow occurs within or at the end of the channel.

For the first type of lateral spillway channel, the starting point for the numerical integration procedure is at the channel discharge point where the water surface elevation is established by a downstream control point. However, for the second type of channel, the numerical integration is started with critical depth.

Equation (3-38) may be employed to compute the critical depth for lateral spillway channels with critical flow in the channel.

$$\left[\frac{gA^3}{q^2B}\right]^{1/2} - \left[\frac{8q^2}{gB^2\left[S - \dfrac{14.5n^2P^{4/3}}{A^{1/3}B}\right]^3}\right] = 0 \qquad (3\text{-}38)$$

where q = discharge per unit length of channel, i.e., Q/L, ft^3/s-ft

n = Manning's roughness coefficient

S = slope of channel bottom, ft/ft

For a trapezoidal channel the coefficients A, B, and P are given by the relationships

$$A = Wy_c + Ky_c^2 \tag{3-39}$$

$$B = W + 2Ky_c \tag{3-40}$$

$$P = W + 2y_c\sqrt{K^2 + 1} \tag{3-41}$$

where the slope of the side wall (rise divided by run) is $1/K$.

If the channel cross section is rectangular, Eq. (3-39), (3-40), and (3-41) reduce to

$$A = Wy_c \tag{3-42}$$

$$B = W \tag{3-43}$$

$$P = W + 2y_c \tag{3-44}$$

Equation (3-38) may be solved by trial and error for y_c. Once y_c is determined, the location, X, of the critical section may be found from the following equation:

$$X_c = \left[\frac{gA^3}{q^2B}\right]^{1/2} \tag{3-45}$$

provided $X_c < L$. Equations (3-38) and (3-45) are only applicable when the channel is long enough for critical depth to occur. In most cases in treatment plant design the most workable approach is to assume y_c occurs at the brink, according to Eq. 3-16.

Once the critical depth and the location of the critical section have been determined, the numerical integration procedure can begin. The computations proceed upstream from the control section, where subcritical flow exists, and downstream from the control section, where supercritical flow exists. The calculation procedure for the lateral spillway channel with free overfall is as follows:

1. Assume critical depth occurs at the brink and compute y_c from Eq. (3-18).

2. Establish a convenient incremental distance, ΔX, to facilitate computations. The value can be a constant or it may vary from step to step.

3. Assume a value for $\Delta y'$ for the first section and compute Δy from Eq. (3-36).

4. Once a value for Δy has been obtained, the one unknown depth for the section under consideration is computed from Eq. (3-35).

5. Determine the discharge at the upstream and downstream ends of the section of interest.

6. Determine the velocity of flow at the upstream and downstream ends of the section of interest.

7. Calculate $\Delta y'$ from Eq. (3-37).

8. Compare the $\Delta y'$ value computed in Step 7 to the value assumed in Step 3. If they agree to within some specified error criterion, the $\Delta y'$ value is correct. If they do not agree to within some specified error criterion, repeat Step 4 through Step 8 with a different assumed $\Delta y'$ value.

9. Repeat Step 3 through Step 8 until the farthest upstream section is reached. The $\Delta y'$ value for this section cannot be computed. According to Chow (1959), it is *assumed* to be *twice* the velocity head at the upstream side of the next to last upstream section.

EXAMPLE PROBLEM 3-4:   Establish the water surface profile for a lateral spillway channel with free overall if the channel is 24 ft long and 1.5 ft wide. The channel receives a total flow of 6.1 cfs evenly distributed along the entire channel length. The channel has a slope of 0.001 and a roughness coefficient, n, of 0.013.

Solution:

1. Assume that critical depth occurs at the brink and compute its value from Eq. (3-18).

$$y_c = \left[\frac{Q^2}{gW^2}\right]^{1/3}$$
$$= \left[\frac{(6.1)^2}{(32.2)(1.5)^2}\right]^{1/3}$$
$$= \mathbf{0.801\ ft}$$

2. The problem is solved by beginning with critical depth and, by applying Eqs. (3-36) and (3-37), moving upstream until the upstream end of the channel is reached. Calculations for a step size of 1.0 ft are summarized in the table on page 101. In this table DYPA represents the assumed value for $\Delta y'$ when DYPA and DYPC were within a relative error criterion of 1.0×10^{-1}, i.e., when

$$\left|\frac{\text{DYPA} - \text{DYPC}}{\text{DYPC}}\right| < 0.1$$

The $\Delta y'$ value for the farthest upstream section cannot be computed from Eq. 3-37 because Q_1 and V_1 are zero. Normally, the $\Delta y'$ value for this section is assumed to be twice the velocity head at the upstream side of the next to last upstream section. Hence, for station 1.00

$$\Delta y' \simeq 2\left[\frac{(0.1262)^2}{2(32.3)}\right]$$
$$\simeq 0.0005$$

The depth of flow at the upstream end of the channel can be computed from Eqs. (3-35) and (3-36).

An increment of 0.0005 units was used to adjust each assumed DYPA value.

y_1 (ft)	y_2 (ft)	Q_1 (ft³/s)	Q_2 (ft³/s)	V_1 (ft/s)	V_2 (ft/s)	DYPA (ft)	DYPC (ft)	Downstream station (ft)
0.959	0.801	5.846	6.100	4.062	5.077	0.159	0.177	24.00
1.026	0.959	5.592	5.846	3.631	4.062	0.068	0.075	23.00
1.075	1.026	5.337	5.592	3.310	3.631	0.049	0.055	22.00
1.113	1.075	5.083	5.337	3.043	3.310	0.039	0.044	21.00
1.146	1.113	4.829	5.083	2.809	3.043	0.033	0.037	20.00
1.173	1.146	4.575	4.829	2.599	2.809	0.028	0.031	19.00
1.197	1.173	4.321	4.575	2.405	2.599	0.025	0.027	18.00
1.218	1.197	4.067	4.321	2.225	2.405	0.022	0.024	17.00
1.237	1.218	3.812	4.067	2.055	2.225	0.019	0.021	16.00
1.253	1.237	3.558	3.812	1.892	2.055	0.017	0.019	15.00
1.268	1.253	3.304	3.558	1.737	1.892	0.015	0.017	14.00
1.281	1.268	3.050	3.304	1.587	1.737	0.014	0.015	13.00
1.292	1.281	2.796	3.050	1.442	1.587	0.012	0.013	12.00
1.302	1.292	2.542	2.796	1.301	1.442	0.011	0.012	11.00
1.311	1.302	2.287	2.542	1.163	1.301	0.009	0.010	10.00
1.318	1.311	2.033	2.287	1.028	1.163	0.008	0.009	9.00
1.325	1.318	1.779	2.033	0.895	1.028	0.007	0.008	8.00
1.330	1.325	1.525	1.779	0.764	0.895	0.006	0.007	7.00
1.335	1.330	1.271	1.525	0.635	0.764	0.005	0.006	6.00
1.338	1.335	1.017	1.271	0.506	0.635	0.004	0.005	5.00
1.341	1.338	0.762	1.017	0.379	0.506	0.003	0.003	4.00
1.342	1.341	0.508	0.762	0.252	0.379	0.002	0.002	3.00
1.343	1.342	0.254	0.508	0.126	0.252	0.001	0.001	2.00
1.342	1.343	—	0.254	—	0.126	—	0.000	1.00

$$\Delta y' = -\Delta y + S\Delta X \qquad \textbf{(3-36)}$$

$$0.0005 = -\Delta y + (0.001)(1)$$

$$\Delta y = 0.0005$$

Since

$$\Delta y = y_2 - y_1 \qquad \textbf{(3-35)}$$

$$y_1 = y_2 - \Delta y$$

$$= 1.3430 - 0.0005$$

$$y_1 = 1.3425$$

See Appendix I for the Fortran computer code for this problem.

In a discussion presented by Thomas (1940), the writer notes that the *arbitrary location* of the critical section obscures the refinement of Eq. (3-37) given by the inclusion of the friction term. As a result, he recommends that the friction term be neglected and further states that "from the standpoint of the designer, it usually suffices to know what depth of water to expect in the upper end of the channel under a particular set of conditions. It is seldom necessary in conventional design to compute the entire profile of the water surface." By neglecting channel friction and assuming that the shape of the water surface approximates a parabola, Thomas (1940) developed Eq. (3-46) (see Fig. 3-13).

$$y_u = \left[2(y_c)^2 + \left(y_c - \frac{SX_c}{3} \right)^2 \right]^{1/2} - \frac{2SX_c}{3} \tag{3-46}$$

where y_u = depth at the upstream end of the channel, ft

X_c = distance from upstream end of the channel to the section where critical depth occurs, ft

y_c = critical depth, ft

Experimental data indicated that little error was introduced if X_c was taken as the total length of the channel, L, and critical depth was based on the total discharge, i.e.,

$$y_c = \left[\frac{Q^2}{gW^2} \right]^{1/3}$$

for a rectangular channel. Hence. Eq. (3-46) may be approximated as

$$y_u = \left[2(y_c)^2 + \left(y_c - \frac{SL}{3} \right)^2 \right]^{1/2} - \frac{2SL}{3} \tag{3-47}$$

For the case of zero slope (S = 0), Eq. (3-47) reduces to the form

$$y_u = 1.73\, y_c \tag{3-48}$$

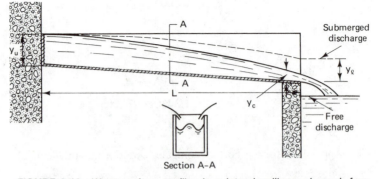

Section A-A

FIGURE 3-13 Water-surface profiles in a lateral spillway channel, from Water Purification and Wastewater Treatment and Disposal, Vol. 2, by Fair, Geyer, and Okun, 1971. Used with permission of John Wiley & Sons, Inc.

For the situation where the depth of flow at the lower end of the channel is fixed at upper stage by a downstream control point (see submerged discharge, Fig. 3-13), Thomas (1940) presents Eq. (3-49) for the depth of flow at the upstream end of the channel.

$$y_u = \left[\frac{2(y_c)^3}{y_l} + \left(y_l - \frac{SL}{3} \right)^2 \right]^{1/2} - \frac{2SL}{3} \tag{3-49}$$

where y_l = the measured or expected depth of water at the lower end of the channel, ft

For the case of zero slope ($S = 0$), Eq. (3-49) reduces to

$$y_u = \left[\frac{2(y_c)^3 + (y_l)^3}{y_l} \right]^{1/2} \tag{3-50}$$

It should be noted that Eq. (3-49) reduces to Eq. (3-47) when y_l equals y_c.

For many situations encountered in treatment plant design, outflow from a component such as a sedimentation basin is controlled by a series of evenly spaced weirs discharging into a lateral spillway channel. In this case Eq. (3-49) may be written as

$$y_u = \left[\frac{2\left\{ \left[\frac{(mq)^2}{gW^2} \right]^{1/3} \right\}^3}{y_l} + \left(y_l - \frac{SL}{3} \right)^2 \right]^{1/2} - 2\left(\frac{SL}{3} \right) \tag{3-51}$$

where m represents the number of weirs between the upstream and downstream end of the channel and q represents the discharge per weir. When the channel is placed on a zero slope, Eq. (3-51) reduces to the form

$$y_u = \left[\frac{2(mq)^2}{gW^2 y_l} + y_l^2 \right]^{1/2} \tag{3-52}$$

EXAMPLE PROBLEM 3-5:   Use the data of Example Problem 3-4 and apply the simplified equations developed by Thomas to compute the water depth at the upstream end of the channel. Compare this value with the upstream depth computed in Example Problem 3-4.

Solution: Compute y from Eq. (3-47) using the critical depth determined in Example Problem 3-4 of 0.801 ft.

$$y_u = \left[2(y_c)^2 + \left(y_c - \frac{SL}{3} \right)^2 \right]^{1/2} - 2\left(\frac{SL}{3} \right)$$

$$= \left[2(0.801)^2 + \left(0.801 - \frac{(0.001)(24)}{3} \right)^2 \right]^{1/2} - \frac{(2)(0.001)(24)}{3}$$

$$= \textbf{1.3668 ft}$$

A value of 1.3425 ft was determined in Example Problem 3-4. Thus, the simplified equations of Thomas give results that compare with the numerical integration procedure. However, it must be remembered that if the channel is rather long, friction losses will become significant and must be considered.

Side-Discharge Weirs

Side-discharge weirs are commonly used at wastewater treatment plants to bypass excess wet weather flow or take off flow peaks for side-line equalization. Henderson (1966) has presented three different flow profiles which may exist at side-discharge weirs. These are shown in Fig. 3-14.

If it is assumed that the specific energy, E, is a constant over the length of the weir and if it is also assumed that the outflow–head relationship for side weirs has the form

$$q = C(H)^{3/2} \tag{3-53}$$

where q = discharge over weir per unit length of weir, ft^2/s

H = head over the weir crest, ft

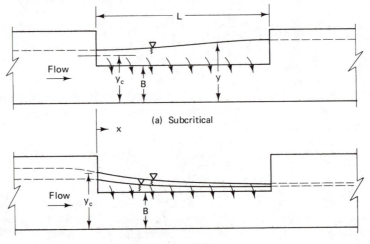

(a) Subcritical

(b) Supercritical

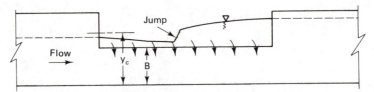

(c) Mixed profile

FIGURE 3-14 *Typical flow profiles at side-discharge weirs, after Henderson (1966).*

C = weir constant, which is often taken as 4.1 for side weirs (Ackers, 1957)

For a *rectangular channel* deMarchi (1934) has shown that the discharge over the weir per unit length of weir is equal to

$$-\frac{dQ}{dX} = C_1\sqrt{2g}\,(y - B)^{3/2} \tag{3-54}$$

where Q = discharge in the main channel, ft^3/sec

$C = C_1\sqrt{2g}$

B = height of weir, ft

The negative sign results because Q decreases along the weir length. Henderson (1966) notes that at any distance X along the weir, the flow in the channel is given by

$$Q_x = Wy\sqrt{2g(E - y)} \tag{3-55}$$

where Q_x is the flow in the channel at the specified distance X, W is the width of the channel, and E is the specific energy.

Henderson (1966) also shows that the change in the depth of flow in the channel with weir length is

$$\frac{dy}{dX} = \frac{2C_1}{W}\,\frac{[(E - y)(y - B)^3]^{1/2}}{(3y - 2E)} \tag{3-56}$$

which can be integrated to give

$$\frac{XC_1}{W} = \frac{2E - 3B}{E - B}\sqrt{\frac{E - y}{y - B}} - 3\,\sin^{-1}\sqrt{\frac{E - y}{E - B}} + \text{constant of integration} \tag{3-57}$$

The design of a side-discharge weir is illustrated in Example Problem 3-6.

EXAMPLE PROBLEM 3-6: Determine the length of side weir required if it is desired that the weir begin discharging when the channel flow is 20 cfs and that when the channel flow is 30 cfs, the weir discharge is 5 cfs. The channel is 6.0 ft wide, has a slope of 0.001 and a roughness coefficient of 0.014.

Solution:

1. Determine the height of the weir, B. The height of the weir has the same numerical value as the normal depth in the channel when $Q = 20$ cfs. Hence, Manning's equation may be used for this calculation.

$$Q = A\frac{1.49}{n} R^{2/3} S^{1/2}$$

$$Q = (Wy_o)\frac{1.49}{n} \left[\frac{Wy_o}{2y_o + W}\right]^{2/3} S^{1/2}$$

$$20 = (6y_o)\frac{1.49}{0.014} \left[\frac{6y_o}{2y_o + 6}\right]^{2/3} (0.001)^{1/2}$$

Solving this relationship by trial and error for y_o gives

$$y_o = B = \textbf{1.128 ft}$$

2. Determine the depth of flow at the downstream end of the weir just after discharge is completed, i.e., when $Q = 25$ cfs.

$$Q = (Wy)\frac{1.49}{n} \left[\frac{Wy}{2y + W}\right]^{2/3} S^{1/2}$$

$$25 = (6y)\frac{1.49}{0.014} \left[\frac{6y}{2y + 6}\right]^{2/3} (0.001)^{1/2}$$

Solving by trial and error for y gives

$$y = \textbf{1.315 ft}$$

3. Calculate the specific energy at the downstream end of the weir where the channel discharge is 25 cfs.

$$E = y + \frac{Q}{2gW^2y^2}$$

$$= 1.315 + \frac{(25)^2}{2(32.2)(6)^2(1.315)^2}$$

$$E = \textbf{1.471 ft}$$

4. Determine the depth of flow at the upstream end of the weir where the channel discharge is 30 cfs. Assuming E is constant along the length of the weir,

$$E = y + \frac{Q^2}{2gW^2y^2}$$

$$1.471 = y + \frac{(30)^2}{2(32.2)(6)^2(y)^2}$$

Solving by trial and error for y gives

$$y = \textbf{1.205 ft}$$

5. Compute the length of the weir by solving Eq. (3-57) between the limits shown on the sketch below.

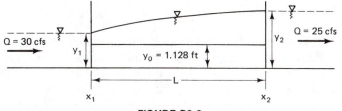

FIGURE P3-6

$$\left.\frac{XC_1}{W}\right|_{x_1}^{x_2} = \left.\frac{XC_1}{W}\right|_0^L = \left\{\frac{2E - 3B}{E - B} \sqrt{\frac{E - y}{y - B}} - 3 \sin^{-1} \sqrt{\frac{E - y}{E - B}}\right\}\Bigg|_{y_1}^{y_2}$$

$$L = \left(\frac{6}{4.1/\sqrt{164.4}}\right)\{[(-1.177) - (2.220)] \\ - [(-2.395) - (3.23)]\}$$

$$L = \mathbf{26.16\ ft}$$

Subramanya and Awasthy (1972) proposed a slightly different form of the de-Marchi equation than that shown by Eq. (3-54). These investigators give Eq. (3-58) for the discharge over the side weir per unit length of weir.

$$q_s = -\frac{dQ}{dX} = \frac{2}{3}(C_w)\sqrt{2g}\,(y - B)^{3/2} \qquad \text{(3-58)}$$

where q_s = discharge over the side weir per unit length of weir, ft²/s
           C_W = discharge coefficient

If the length of the weir is L, then the total discharge over the weir is

$$q_w = \frac{2}{3}(L)(C_w)\sqrt{2g}\,(y - B)^{3/2} \qquad \text{(3-59)}$$

where q_w = total discharge over the side weir, ft³/s

The discharge coefficient C_w is not a constant but varies with the mean velocity of flow in the *upstream* main channel. According to Subramanya and Awasthy (1972), C_w may be estimated from the relationship

$$C_w = 0.611\sqrt{1 - \left[\frac{3(F_u)^2}{(F_u)^2 + 2}\right]} \qquad \text{(3-60)}$$

where F_u represents the Froude number in the upstream main channel and may be computed from Eq. (3-25).

Distribution Channels

In water and wastewater treatment plants, open channels rather than manifolds are often used to distribute incoming flow to parallel treatment units, such as aeration tanks, sedimentation basins, filters, and flocculation basins. Such channels are referred to as *distribution channels*. Flow control in these channels is normally achieved by rectangular weirs, V-notch weirs, and submerged orifices. These control devices must be designed so that the incoming flow is evenly distributed to the process units. Thus, the analysis of open channel distribution conduits is an important problem faced by many engineers.

Chao and Trussell (1980) have outlined a method for analyzing flow distribution in a rectangular, prismatic, horizontal channel when flow control is achieved by either rectangular side weirs or submerged orifices. These workers proposed a step method wherein the flow characteristics of each weir or orifice is determined by proceeding step by step from the downstream end of the distribution channel to the upstream end where the flow enters the channel. The procedure proposed by Chao and Trussell (1980) for determining the flow distribution when rectangular side weirs are used for flow control is outlined below:

1. Establish the number of weirs to be used, the length of each weir, the height of each weir above the channel bottom, the width of the channel, and the total inflow.

2. Start at the downstream end of the channel and assume a flow over the final weir.

$$(q_w)_{fa} = \frac{\text{total inflow}}{\text{number of weirs}}$$

3. With the assumed flow from Step 2, estimate a depth of flow over the final weir and then adjust the flow by using the information obtained from the weir discharge equation [Eq. (3-59)].

    (a) Assume a depth of flow at the upstream end of the final weir, for example,

$$(y_u)_f = B + 0.01$$

    (b) Estimate the velocity in the channel at the upstream end of the final weir. Since all the flow at the upstream end of the final weir goes over the weir, the flow in the channel at the upstream end of the weir is the same as the flow over the weir.

$$(V_u)_f = \frac{(q_w)_{fa}}{W(y_u)_f}$$

(c) Compute the Froude number at the upstream end of the final weir.

$$(F_u)_f = \frac{(V_u)_f}{\sqrt{g(y_u)_f}}$$

(d) Estimate the weir coefficient, C_w, for the final weir by applying Eq. (3-60).

$$C_w = 0.611 \sqrt{1 - \left[\frac{3(F_u)_f^2}{(F_u)_f^2 + 2}\right]}$$

(e) Calculate the discharge over the final weir by applying Eq. (3-59).

$$(q_w)_{fc} = \frac{2}{3}(L)(C_w)\sqrt{2g}\,[(y_u)_f - B]^{3/2}$$

For the final weir, the depth of flow is assumed to be constant over the length of the weir.

(f) Compare the assumed weir discharge and the calculated weir discharge.

$$\Delta q = |(q_w)_{fc} - (q_w)_{fa}|$$

If the flow increment, Δq, is greater than some preestablished criterion (e.g., 0.01), increase y by some preestablished criterion (e.g., 0.005) and repeat Steps 3(b) through (f). If Δq is less than the preestablished criterion, move upstream to the next weir.

(g) Compute the specific energy at the upstream end of the final weir. Since it is assumed that E is constant throughout the distribution channel, this value will be used in all upstream computations.

$$E = (y_u)_f + \frac{[(q_w)_{fc}/W(y_u)_f]^2}{2g}$$

4. Initially, assume that the flow over the next weir of interest is the same as that calculated over the nearest downstream weir.

$$(q_w)_k = (q_w)_{k+1}$$

(a) Assume that the depth of flow at the upstream end of the nearest downstream weir is the same as the depth of flow at the downstream end of the weir of interest.

$$(y_D)_k = (y_u)_{k+1}$$

(b) Estimate the flow in the main channel just upstream from the weir of interest

$$(Q_u)_k = (q_w)_k + \sum_{i=k+1}^{i=N} (q_w)_i$$

where N represents the total number of weirs in the channel.

(c)   Compute the depth of flow at the upstream end of the weir of interest by applying Eq. (3-55).

$$(Q_u)_k = W(y_u)_k \sqrt{2g[E - (y_u)_k]}$$

Solve for $(y_u)_k$ by trial and error.

(d)   Determine the average depth of flow along the length of the weir.

$$(y_a)_k = \frac{(y_u)_k + (y_D)_k}{2}$$

(e)   Compute the velocity in the channel at the upstream end of the weir of interest.

$$(V_u)_k = \frac{(Q_u)_k}{W(y_u)_k}$$

(f)   Calculate the Froude number at the upstream end of the weir of interest.

$$(F_u)_k = \frac{(V_u)_k}{\sqrt{g(y_u)_k}}$$

(g)   Estimate the weir coefficient, C_w, for the weir of interest, from Eq. (3-60).

$$C_w = 0.611 \sqrt{1 - \left[\frac{3(F_u)_k^2}{(F_u)_k^2 + 2}\right]}$$

(h)   Calculate the discharge over the weir of interest from Eq. (3-59).

$$(q_w)_{kc} = \frac{2}{3}(L)(C_w)\sqrt{2g}[(y_a)_k - B]^{3/2}$$

(i)   Compare the assumed weir discharge and the calculated weir discharge.

$$\Delta q = |(q_w)_{kc} - (q_w)_k|$$

If the flow increment, Δq, is greater than some preestablished criterion (e.g., 0.01), increase $(q_w)_k$ by some preestablished criterion (e.g., 0.005) and repeat Steps 4(a) through (i). If Δq is less than the preestablished criterion, move upstream to the next weir.

5.   After the discharge over each weir has been estimated, sum the individual calculated discharges.

$$QT = \sum_{i=1}^{i=N} (q_w)_i$$

6. Compare the total calculated discharge, QT, to the actual inflow, Q.

$$Q = |Q - QT|$$

If the total flow increment, ΔQ, is greater than some preestablished criterion (e.g., $N \times 0.05$) increase or decrease $(q_w)_{fa}$ by some preestablished criterion (if QT is greater than Q, decrease $(q_w)_{fa}$; however, if QT is less than Q, increase $(q_w)_{fa}$ and repeat Steps 3 through 6. If ΔQ is less than the preestablished criterion, the actual flow distribution has been established.

To illustrate the calculation procedure required to estimate the flow distribution in a horizontal rectangular distribution channel with rectangular side weirs for flow control, Chao and Trussell (1980) solved the problem presented as Example Problem 3-7.

EXAMPLE PROBLEM 3-7:   A water treatment plant is designed to handle a flow of 15 million gpd. The flow enters from one end of the influent channel and is distributed to two sedimentation basins through six side weirs. The channel is 48 ft long and 4 ft wide. The side weirs are 4 ft wide and placed 4 ft apart as shown in Fig. 3-15. Estimate the discharge distribution through all weirs.

Solution: Following the computational procedure previously discussed, the flow distribution shown below is obtained (see Appendix 1 for the Fortran computer code for this problem):

Weir no.	Flow through weir (cfs)	Upstream depth (ft)
6	4.418	2.485
5	4.336	2.475
4	4.131	2.460
3	3.822	2.435
2	3.456	2.410
1	3.073	2.380

The weir discharge values indicate a significant variation in discharge along the channel (the discharge over weir 6 is approximately 44% greater than the discharge over weir 1).

A number of modifications to the basic channel design may be made which will

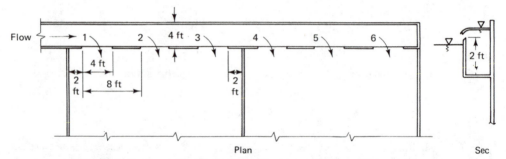

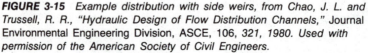

FIGURE 3-15 *Example distribution with side weirs, from Chao, J. L. and Trussell, R. R., "Hydraulic Design of Flow Distribution Channels," Journal Environmental Engineering Division, ASCE, 106, 321, 1980. Used with permission of the American Society of Civil Engineers.*

correct the inequitable flow distribution. Three of those proposed by Chao and Trussell (1980) are presented below:

1. Vary the weir elevations so that weir 1 is the lowest and weir 6 is the highest. The flow distribution for this situation is as follows (the Fortran computer code is given in Appendix 1):

Weir no.	Flow through weir (cfs)	Weir elevation (ft)
6	3.908	2.00
5	3.986	1.99
4	3.942	1.98
3	3.895	1.96
2	3.739	1.94
1	3.754	1.90

In this case the largest flow deviation is only about 7%.

2. Widen the channel. The engineer must consider the possibility of solids settling out in the channel if this modification is used.

3. Taper the channel to keep the Froude number nearly constant (see Fig. 3-16). The flow distribution for this situation is as follows (the Fortran computer code is given in Appendix I):

Weir no.	Flow through weir (cfs)	Upstream depth (ft)
6	3.848	2.442
5	3.879	2.445
4	3.899	2.445
3	3.899	2.445
2	3.899	2.445
1	3.899	2.445

In this case the largest flow deviation is only about 1%.

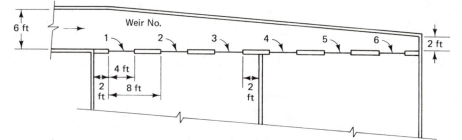

FIGURE 3-16 *Tapered influent channel with side-discharge weirs, from Chao, J. L. and Trussell, R. R., "Hydraulic Design of Flow Distribution Channels," Journal Environmental Engineering division, ASCE, 106, 321, 1980. Used with permission of the American Society of Civil Engineers.*

4. Taper the channel coupled with minor adjustments in the elevation of each weir (the elevation of weir 1 must be slightly less than the elevation of weir 6).

If submerged orifices are used for flow control (see Fig. 3-17), the analysis for flow distribution is similar to that given for side weirs. The primary difference is in the discharge equation for the control device. In the case of submerged orifices the discharge through the device is given by Eq. (3-61).

$$q_o = C_D a \sqrt{2g(\Delta E)} \tag{3-61}$$

where
$\quad q_o$ = discharge through the orifice, cfs

$\quad a$ = cross-sectional area of the orifice, ft^2

$\quad g$ = gravity constant, ft/s^2

$\quad \Delta E$ = the difference in total head across the orifice, ft

$\quad C_D$ = discharge coefficient

The value of the discharge coefficient depends to a large degree on the geometrical

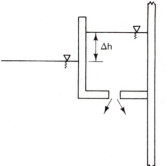

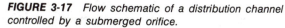

FIGURE 3-17 *Flow schematic of a distribution channel controlled by a submerged orifice.*

characteristics of the orifice. For a circular orifice with a rounded entrance, C_D is given by the expression

$$C_D = 0.975 \left[1 - \frac{V_u^2}{2gE} \right]^{3/8} \tag{3-62}$$

where E represents the total specific energy at the orifice of interest. If the entrance to a circular orifice is sharp edged, then

$$C_D = 0.63 - 0.58 \left[\frac{V_u^2}{2gE} \right] \tag{3-63}$$

For a rectangular sharp-edged orifice, C_D is given by Eq. (3-64).

$$C_D = 0.611 - 0.58 \left[\frac{V_u^2}{2gE} \right] \tag{3-64}$$

Chao and Trussell (1980) indicated that in a distribution channel the velocity head is usually small and can be neglected. In this case Eq. (3-64) reduces to the form

$$C_D = 0.611 - 0.29(F_u)^2 \tag{3-65}$$

The computation procedure for determining the flow distribution when rectangular sharp-edged submerged orifices are used for flow control is presented below:

1. Establish the elevation of channel bottom and the elevation of the exit control weir for the unit of interest (where more than one unit is involved assume the elevation of the exit weir is the same for all units).

    EL_1 = elevation of the distribution channel bottom, ft

    EL_2 = elevation of the exit control weir for the process units, ft

2. Establish the number of orifices to be used, the spacing of the orifices, the size of the orifices, the width of the distribution channel, and the total inflow.

3. Assume equal flow distribution to all units and compute the head on the exit weir. For sharpcrested weirs the appropriate expression is

$$H = \frac{1}{(C_w)^{2/3}(L)^{2/3}} (Q_p)^{2/3}$$

where H = head on weir, ft

C_w = weir coefficient where a value of 3.33 is normally assumed for sharpcrested weirs

L = length of weir, ft

Q_p = flow through the process unit, ft³/s

Initial Q_p is assumed to be given by Q/NP where Q represents the total inflow to the channel and NP represents the number of process units.

4. Start at the downstream end of the distribution channel and assume a flow through the final orifice.

$$(q_o)_{fa} = \frac{\text{total inflow}}{\text{number of orifices}}$$

5. With the assumed flow from Step 4, estimate a depth of flow at the upstream side of the final orifice and then adjust the flow by using the information obtained from the orifice discharge equation [Eq. (3-61)].

(a) Assume a depth of flow at the upstream side of the final orifice, for example,

$$(y)_f = 0.01 \text{ ft}$$

(b) Estimate the velocity in the channel at the upstream side of the final orifice. Since all the flow at the upstream side of the final orifice goes through the orifice, the flow in the channel at the upstream side of the orifice is the same as the flow through the orifice.

$$(V_u)_f = \frac{(q_o)_{fa}}{W(y)_f}$$

(c) Compute the Froude number at the upstream side of the final orifice.

$$(F_u)_f = \frac{(V_u)_f}{\sqrt{g(y)_f}}$$

(d) Estimate the orifice coefficient, C_D, for the final orifice by applying Eq. (3-65).

$$C_D = 0.611 - 0.29 (F_u)_f^2$$

(e) Calculate the discharge through the final orifice by applying Eq. (3-61).

$$(q_o)_{fc} = C_D a \sqrt{2g(\Delta E)}$$

where $\Delta E = \left[EL_1 + (y)_f + \frac{(V_u)_f^2}{2g} \right] - [EL_2 + H]$

It is assumed that the velocity head is zero in the process unit.

(f) Compare the assumed orifice discharge and the calculated orifice discharge.

$$\Delta q = |(q_o)_{fc} - (q_o)_{fa}|$$

If the flow increment, Δq, is greater than some preestablished criterion (e.g., 0.01), increase $(y)_f$ by some preestablished criterion (e.g., 0.005) and repeat Steps 5(b) through (f). If Δq is less than the preestablished criterion, move upstream to the next orifice.

(g) Compute the specific energy at the upstream side of the final orifice. Since it is assumed that E is constant throughout the distribution channel, this value will be used in all upstream calculations.

$$E = (y)_f + \frac{(V_u)_f^2}{2g}$$

6. Initially, assume that the flow through the next orifice of interest is the same as that calculated through the nearest downstream orifice.

$$(q_o)_k = (q_o)_{k+1}$$

(a) Assume that the depth of flow at the upstream side of the nearest downstream orifice is the same as the depth of flow at the upstream side of the orifice of interest.

$$(y)_k = (y)_{k+1}$$

(b) Estimate the flow in the main channel just upstream from the orifice of interest.

$$(Q_u)_k = (q_o)_k + \sum_{i=k+1}^{i=N} (q_o)_i$$

where N represents the total number of orifices in the channel.

(c) Compute the depth of flow at the upstream side of the orifice of interest by applying Eq. (3-55).

$$(Q_u)_k = W(y)_k \sqrt{2g[E - (y)_k]}$$

Solve for $(y)_k$ by trial and error.

(d) Compute the velocity in the channel at the upstream side of the orifice of interest.

$$(V_u)_k = \frac{(Q_u)_k}{W(y)_k}$$

(e) Calculate the Froude number at the upstream side of the orifice of interest.

$$(F_u)_k = \frac{(V_u)_k}{\sqrt{g(y)_k}}$$

(f) Estimate the orifice coefficient, C_D, for the orifice of interest by applying Eq. (3-65).

$$C_D = 0.611 - 0.29(F_u)_k^2$$

(g) Calculate the discharge through the orifice of interest by applying Eq. (3-61).

$$(q_o)_{kc} = C_D a \sqrt{2g(\Delta E)}$$

where $\Delta E = \left[EL_1 + (y)_k + \dfrac{(V_u)_k^2}{2g} \right] - [EL_2 + H]$

(h) Compare the assumed orifice discharge and the calculated orifice discharge.

$$\Delta q = |(q_o)_{kc} - (q_o)_k|$$

If the flow increment, Δq, is greater than some preestablished criterion (e.g., 0.01), increase $(q_o)_k$ by some preestablished criterion (e.g., 0.005) and repeat Steps 6(a) through (h). If Δq is less than the preestablished criterion, move upstream to the next orifice.

7. After the discharge through each orifice has been estimated, sum the individual calculated discharges.

$$QT = \sum_{i=1}^{i=N} (q_o)_i$$

8. Compare the total calculated discharge, QT, to the actual inflow, Q.

$$\Delta Q = |Q - QT|$$

If the total flow increment, ΔQ, is greater than some preestablished criterion (e.g., $N \times 0.01$), increase or decrease $(q_o)_{fa}$ by some preestablished criterion (if QT is greater than Q, decrease $(q_o)_{fa}$; however, if QT is less than Q, increase $(q_o)_{fa}$ and proceed to Step 9. If ΔQ is less than the preestablished criterion, the actual flow distribution has been established.

9. Use the estimated orifice flows to obtain the new flow distribution to each process unit. For example,

Flow through unit 1 $= (q_o)_1 + (q_o)_2 + (q_o)_3 = (Q_p)_1$

Flow through unit 2 $= (q_o)_4 + (q_o)_5 + (q_o)_6 = (Q_p)_2$

Repeat Steps 3 through 8. Example Problem 3-8 is presented to illustrate the calculation procedure required to estimate the flow distribution in a horizontal rectangular distribution channel with submerged orifices for flow control.

EXAMPLE PROBLEM 3-8:   A water treatment plant is designed to handle a flow of 15 million gpd. The flow enters from one end of the influent channel and is distributed to two sedimentation basins through eight submerged orifices located in the bottom of the channel. The channel is 48 ft long and 4 ft wide, and each basin is fed by four orifices. The flow from each sedimentation basin is controlled by a sharp-crested weir 24 ft in length which is located at an elevation of 300.00 ft. The channel bottom is horizontal and is located at an elevation of 299.50 ft. The orifices are evenly spaced along the channel, and each is square with a width of 0.75 ft. Estimate the discharge distribution through all orifices.

Solution: Following the computational procedure previously discussed the flow distribution shown below is obtained (see Appendix I for the Fortran computer code for this problem):

Orifice no.	Flow through orifice (cfs)
8	2.982
7	2.973
6	2.959
5	2.939
4	2.919
3	2.883
2	2.837
1	2.780

In this case the discharge through orifice 8 is only 7.2% greater than the discharge through orifice 1. The total flow through basin 2 is 11.85 million gpd compared to a flow of 11.42 million gpd through basin 1. If a better flow distribution is desired, a smaller orifice size should be used.

3.3 MINOR LOSSES

Minor losses in open channels occur at bends, gate recesses, contractions, enlargements, etc. Since the length of most of the open channels used in water and wastewater treatment plants is quite small, friction losses in these channels are generally insignificant. The energy loss in these channels is primarily associated with minor losses. Hence, the engineer must be able to estimate such energy losses with some degree of accuracy if a realistic hydraulic profile for the plant is to be constructed.

As was the case for pipe flow, minor losses in *channel bends* and gate recesses are commonly expressed in terms of velocity head using the relationship

$$h_L = K\left(\frac{V^2}{2g}\right) \tag{3-66}$$

where K is referred to as the head loss coefficient and V is the average velocity in the channel before the flow passes through the bend. Many engineers use a value of 0.2 for K for 45° channel bends and gate recesses and a value of 0.3 for K for 90° channel bends.

Plan views of an abrupt channel expansion and abrupt channel contraction are presented in Fig. 3-18. For the case of a channel expansion Henderson (1966) shows that the energy loss between the upstream and downstream section is given by Eq. (3-67).

$$E_1 - E_2 = \Delta E = \frac{V_1^2}{2g}\left[\left(1 - \frac{W_1}{W_2}\right)^2 + \frac{2(F_u)_1^2(W_1)^3(W_2 - W_1)}{(W_2)^4}\right] \qquad \textbf{(3-67)}$$

where

E_1 = specific energy of the upstream section, ft

E_2 = specific energy of the downstream section, ft

V_1 = average velocity of flow in the upstream section, ft/s

W_1 = width of the upstream channel, ft

W_2 = width of the downstream channel, ft

$(F_u)_1$ = Froude number for the upstream section.

g = gravity constant, ft/s^2

Brater and King (1976) proposed that the energy loss for an expansion be computed from the relationship

$$h_L = K_E\left[\frac{V_1^2}{2g} - \frac{V_2^2}{2g}\right] \qquad \textbf{(3-67)}$$

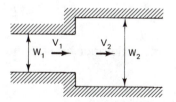

(a) Channel expansion

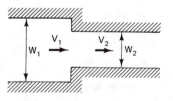

(b) Channel contraction

FIGURE 3-18 *Plan views of abrupt changes in channel geometry.*

Different values of the head loss coefficient shown in this equation are presented in Table 3-1.

Head losses through contractions are less than head losses through expansions. Henderson (1966) indicated that the maximum head loss for square-edged contractions in rectangular channels may be estimated from the relationship

$$(h_L)_{max} \simeq 0.23\left(\frac{V_2^2}{2g}\right) \tag{3-68}$$

If the edges of the contraction are rounded, the maximum head loss may be approximated from

$$(h_L)_{max} \simeq 0.11\left(\frac{V_2^2}{2g}\right) \tag{3-69}$$

Brater and King (1966) presented Eq. (3-70) to describe the head loss through a contraction.

$$h_L = K_c\left[\frac{V_2^2}{2g} - \frac{V_1^2}{2g}\right] \tag{3-70}$$

Values normally used for K_c are presented in Table 3-1.

When there is a transition such as a contraction or expansion in a channel, the elevation of the water surface would change even if there were no energy loss through the transition. The change in the water surface elevation may be determined by applying the Bernoulli equation. For example, consider the expansion shown in Fig. 3-19. Writing the Bernoulli equation across the transition when it is assumed that no energy loss occurs gives

$$\frac{V_1^2}{2g} + y_1 = \frac{V_2^2}{2g} + y_2 \tag{3-71}$$

According to this equation, the change in the water surface elevation between the upstream and downstream sections is

TABLE 3-1 *Typical values for* K_E *and* K_c, *after Brater and King (1966).*

Form of transition	K_E	K_c
Sudden change in area,		
sharp corners	1.0	0.5
Well-designed*	0.1	0.05
Best design value	0.2	0.10

* A well-designed transition is one in which all plane surfaces are connected by tangent curves and a straight line connecting flow lines at the two ends does not make an angle greater than 12.5° with the axis of the channel.

Source: Handbook of Hydraulics, sixth ed. by Brater and King. Used with permission of McGraw-Hill Book Co.

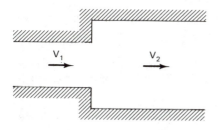

(a) Plan view of channel expansion

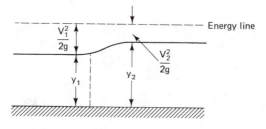

Energy line

(b) Profile of water surface profile through channel expansion

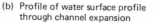

FIGURE 3-19 *Flow schematic of a channel expansion.*

$$y_1 - y_2 = \Delta y = \frac{V_2^2}{2g} - \frac{V_1^2}{2g} \tag{3-72}$$

If an energy loss is assumed to occur across the expansion, the Bernoulli equation has the form

$$\frac{V_1^2}{2g} + y_1 = \frac{V_2^2}{2g} + y_2 + h_L \tag{3-73}$$

and the change in the water surface elevation between the upstream and downstream sections is

$$\Delta y = \frac{V_2^2}{2g} - \frac{V_1^2}{2g} + h_L \tag{3-74}$$

REFERENCES

ACKERS, P., "A Theoretical Consideration of Side Weirs as Stormwater Overflows," *Proceedings, Institute of Civil Engineers,* **120,** 255 (1957).

ALBERTSON, M. L., BARTON, J. R., and SIMONS, D. B., *Fluid Mechanics for Engineers,* Prentice–Hall, Inc., Englewood Cliffs, New Jersey (1960).

BRATER, E. F., and KING, H. W., *Handbook of Hydraulics,* 6th Edition, McGraw–Hill Book Company, New York, New York (1976).

CAMP, T. R., "Lateral Spillway Channels," *Transactions, ASCE,* **105,** 606 (1940).

CAMP, T. R., "Design of Sewers to Facilitate Flow," *Sewage Works Journal,* **18,** 1 (1946).

CAMP, T. R., "Applied Hydraulic Design of Treatment Plants, (Part I)," In *Seminar Papers on Wastewater Treatment and Disposal,* edited by G. M. Reece, Boston Society of Civil Engineers, 231, Boston, Massachusetts (1961).

CHAO, J. L., and TRUSSELL, R. R., "Hydraulic Design of Flow Distribution Channels," *Journal Environmental Engineering Division, ASCE,* **106,** 321 (1980).

CHOW, V. T., *Open-Channel Hydraulics,* McGraw–Hill Book Company, New York, New York (1959).

DAKE, J. M. K., *Essentials of Engineering Hydraulics,* Wiley Interscience, New York, New York (1972).

DAUGHERTY, R. L., and FRANZINI, J. B., *Fluid Mechanics with Engineering Applications,* 7th Edition, McGraw–Hill Book Company, New York, New York (1977).

DEMARCHI, G., "Essay on the Performance of Lateral Weirs," *L'Energia elletrica,* Milan, Italy, Vol. 11, No. 11 (November 1934).

FAIR, G. M., GEYER, J. C., and OKUN, D. A., *Elements of Water Supply and Wastewater Disposal,* 2nd Edition, John Wiley & Sons, New York, New York (1971).

HENDERSON, F. M., *Open Channel Flow,* Macmillan Inc., New York, New York (1966).

SUBRAMANYA, K. and AWASTHY, S. C., "Spatially Varied Flow over Side-Weirs," *Journal Environmental Engineering Division, ASCE,* **98,** 1 (1972).

THOMAS, H. A., Discussion of T. R. Camp's paper, "Lateral Spillway Channels," *Transactions, ASCE,* **105,** 627 (1940).

WEBBER, N. B., *Fluid Mechanics for Civil Engineers,* E. & F. N. Spon Ltd., London, England (1965).

4

FLOW MEASUREMENT
AND HYDRAULIC
CONTROL POINTS

In order to operate a water or wastewater treatment plant properly, it is necessary to know the amount of flow entering the plant and the flow rate of certain process streams within the plant. Several types of flow-measuring devices are available to provide this information including weirs, Parshall flumes, venturi tubes, and magnetic flow meters. Each of these devices is discussed below.

4.1 WEIRS

A weir may be defined as "any regular obstruction over which flow occurs" (Vennard, 1966). Although weirs are extremely simple devices, they provide one of the most accurate methods of measuring open channel flow. However, they should not be used in liquid streams that contain settleable solids, such as raw wastewater, because the solids have a tendency to settle behind the weir. Thus, their use for wastewater flow measurement is generally limited to the measurement of effluent flows that are low in settleable solids. An exception to this would be the use of a proportional weir as a velocity control device for a horizontal, gravity-type, grit chamber.

The most common use of weirs in water and wastewater treatment plants is not for flow measurement, but as effluent structures for units such as aerated grit chambers, clarifiers, aeration basins, and settling basins. When used in this manner, weirs serve as hydraulic control points within the plant as discussed in Sec. 4.6.

The flow over a weir represents a complex hydraulic situation; however, relatively simple equations have been developed to describe the head–discharge relationship. Although weirs are available in various shapes, rectangular and triangular weirs are most frequently encountered.

Rectangular Weirs

Rectangular weirs (sharp-crested weirs) consist of straight, horizontal crests over which liquid flows. The two most important variables affecting the hydraulic profile are the head on a weir (H), and the extent of ventilation beneath the nappe. The head is a function of discharge, but the extent of ventilation can be controlled by controlling the downstream tailwater elevation. If the tailwater elevation permits the weir to discharge freely, [Fig. 4-1(a)], atmospheric pressure exists beneath the nappe; however, if the tailwater elevation is allowed to rise above the weir crest, the profile appears as in Fig. 4-1(b) and the weir is said to be submerged. Standard weir equations are derived for freely discharging weirs, and serious errors will result if they are applied to submerged weirs.

The Francis equation is the most popular equation for describing the head–discharge relationship for freely discharging rectangular weirs. The Francis equation for suppressed weirs was presented as Eq. (1-35) and is given below in a somewhat different form:

$$Q = 3.33 \, LH^{3/2} \tag{4-1}$$

where Q = discharge, ft^3/s

   L = weir length, ft

   H = head on the weir, ft

For contracted rectangular weirs, Fig. 4-2(b), the Francis equation is

$$Q = 3.33(L - 0.1(n)H)H^{3/2} \tag{4-2}$$

where n = number of end contractions

and the other terms are previously defined.

Rectangular weirs are frequently used as effluent structures for aerated grit chambers, aeration basins, and chlorine contact chambers. When used in this manner, the weirs are normally designed for free discharge because, in this mode, they are more

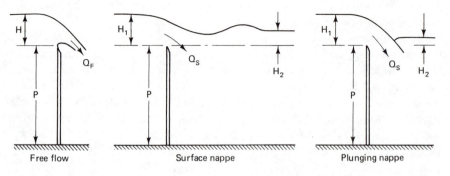

(a) Free discharge   (b) Submerged discharge

FIGURE 4-1 *Discharge profiles for sharp-crested weirs. (a) Free discharge; (b) submerged discharge, after Vennard and Weston (1943).*

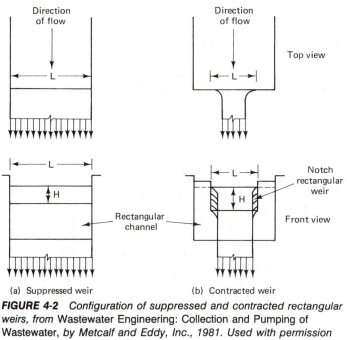

FIGURE 4-2 *Configuration of suppressed and contracted rectangular weirs, from* Wastewater Engineering: Collection and Pumping of Wastewater, *by Metcalf and Eddy, Inc., 1981. Used with permission of McGraw-Hill Book Company.*

precise and their performance is more predictable. One disadvantage of a freely discharging weir, however, is the relatively large amount of head that must be sacrificed to ensure free discharge [see Fig. 4-1(a)]. This head, once lost, cannot be recovered and must be accounted for in the hydraulic design of the plant.

EXAMPLE PROBLEM 4-1: Calculate the head on a suppressed rectangular weir at a flow of 3.0 cfs if the weir is 5 ft long. (see Fig. 4-2.)

Solution: Apply Eq. (4-1) and compute the head on the weir.

$$Q = 3.33(L)H^{3/2}$$

$$H = \left[\frac{Q}{3.33(L)}\right]^{2/3} = \left[\frac{3.0}{3.33(5)}\right]^{2/3}$$

$$H = \mathbf{0.32\ ft}$$

The performance of submerged rectangular weirs cannot be described by the same equations that apply to freely discharging weirs, and they must be recognized as a special situation. Figure 4-3 was developed by Vennard and Weston (1943) to describe the performance of submerged rectangular weirs. Referring to Fig. 4-1, the variables of interest are

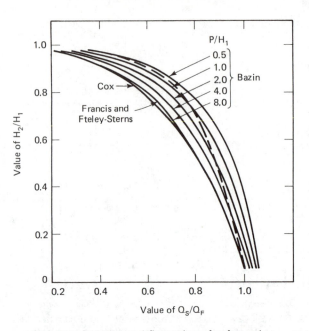

Note: Dotted curve defines regions of surface and plunging nappes. To the right and above this line is region of plunging nappes and that to the left and below region of surface nappes.

FIGURE 4-3 *Head–discharge relationship for submerged weirs, after Vennard and Weston (1943).*

H_1 = upstream head on weir, ft

H_2 = downstream head on weir, ft

Q_s = submerged flow rate, cfs

Q_F = discharge that would have existed with free discharge for a head H_1, cfs

P = distance from channel bottom to weir crest, ft

EXAMPLE PROBLEM 4-2:   A rectangular weir is to be operated as a submerged weir to discharge a flow of 5.0 cfs. The weir is 6 ft long and is to be operated with an upstream head (H_1) of 0.44 ft. Calculate the downstream head (H_2) that should exist to yield the desired conditions. The distance from the channel bottom to the weir crest, P, is 1.76 ft.

Solution:

1. Calculate Q_s using Eq. (4-1).

$$Q_s = 3.33 \, LH^{3/2}$$

$$H = H_1 = 0.44 \text{ ft}$$

$$L = 6.0 \text{ ft}$$

$$Q_s = 3.33(6.0)(0.44)^{3/2}$$

$$Q_s = \mathbf{5.83\ cfs}$$

2. Calculate Q_s/Q_F.

$$\frac{Q_s}{Q_F} = \frac{5.00\ \text{cfs}}{5.83\ \text{cfs}} = \mathbf{0.86}$$

3. Enter Fig. 4-3 with $Q_s/Q_F = 0.86$ and read H_2/H_1.

$$\frac{H_2}{H_1} = \mathbf{0.47}$$

4. Compute H_2

$$H_2 = H_1(0.47)$$

$$= 0.44\ \text{ft}\ (0.47) = \mathbf{0.21\ ft}$$

Generally, submerged weirs should not be used to measure discharge because of the uncertainties associated with the effect of submergence.

Triangular Weirs

A triangular weir is a popular type of flow-measuring device and is frequently used for measuring small flow rates. Although several different notch angles are available, the 90° V-notch weir is the most common. The head–discharge relationship for a freely discharging 90° weir is

$$Q = 2.5\ H^{2.5} \qquad\qquad\qquad \textbf{(4-3)}$$

where Q = discharge, cfs

H = head on the weir, ft

The use of this equation is illustrated in the following example.

EXAMPLE PROBLEM 4-3: Calculate the head on a 90° V-notch weir handling a discharge of 0.75 cfs.

Solution: Apply Eq. (4-3) and compute the head on the weir

$$Q = 2.5\ H^{2.5}$$

$$H = \left[\frac{Q}{2.5}\right]^{1/2.5} = \left[\frac{Q}{2.5}\right]^{0.4}$$

$$H = \left[\frac{0.75}{2.5}\right]^{0.4} = \mathbf{0.62\ ft}$$

Saw-toothed weirs, consisting of many small 90° weirs, are normally used as effluent structures in clarifiers at wastewater treatment plants. In these situations the effluent flow leaving the clarifier must be divided by the total number of notches available on the weir to determine the discharge per notch. Equation (4-3) can then be applied to calculate the head on the weir as illustrated in Example Problem 4-8.

4.2 PARSHALL FLUMES

Parshall flumes are widely used for measuring the flow in open channels and consist of a converging section, a throat section, and a diverging section as shown in Fig. 4-4. The floor of the throat section is inclined downward, and the floor of the diverging section is inclined upward. This geometry causes critical depth to occur near the beginning of the throat section and produces a backwater curve whose depth, measured at point H_a, can be related to discharge. A second gaging point, H_b, is located at the downstream end of the throat section, and the ratio of H_b to H_a defines the submergence of the flume.

The capacity of a Parshall flume is determined by its throat width, which may range from 3 in. up to 50 ft. The relationships between gage reading H_a and discharge for flumes of various sizes are given by the following equations:

Throat width	Equation	
3 in.	$Q = 0.992\,H_a^{1.547}$	(4-4)
6 in.	$Q = 2.06\,H_a^{1.58}$	(4-5)
9 in.	$Q = 3.07\,H_a^{1.53}$	(4-6)
12 in. to 8 ft	$Q = 4W\,H_a^{(1.522W)^{0.26}}$	(4-7)

W	A	$\frac{2}{3}$A	B	C	D	E	F	G	K	N	R	M	P	X	Y	Free-flow capacity Min.	Free-flow capacity Max.
Ft. In.	Ft. In.	Ft. In.	Ft. In.	Ft. In.	Ft. In.	Ft. In.	Ft. In.	Ft. In.	In.	In.	Ft. In.	Ft. In.	Ft. In.	In.	In.	Cfs	Cfs
0 3	1 $6\frac{3}{8}$	1 $\frac{1}{4}$	1 6	0 7	0 $10\frac{3}{16}$	2 0	0 6	1 0	1	$2\frac{1}{4}$	1 4	1 0	2 $6\frac{1}{4}$	1	$1\frac{1}{2}$	0.03	1.9
0 6	2 $\frac{7}{16}$	1 $4\frac{5}{16}$	2 0	1 $3\frac{1}{2}$	1 $3\frac{5}{8}$	2 0	1 0	2 0	3	$4\frac{1}{2}$	1 4	1 0	2 $11\frac{1}{2}$	2	3	0.05	3.9
0 9	2 $10\frac{5}{8}$	1 $11\frac{1}{8}$	2 10	1 3	1 $10\frac{5}{8}$	2 6	1 0	1 6	3	$4\frac{1}{2}$	1 4	1 0	3 $6\frac{1}{2}$	2	3	0.09	8.9
1 0	4 6	3 0	4 $4\frac{7}{8}$	2 0	2 $9\frac{1}{4}$	3 0	2 0	3 0	3	9	1 8	1 3	4 $10\frac{3}{4}$	2	3	0.11	16.1
1 6	4 9	3 2	4 $7\frac{7}{8}$	2 6	3 $4\frac{3}{8}$	3 0	2 0	3 0	3	9	1 8	1 3	5 6	2	3	0.15	24.6
2 0	5 0	3 4	4 $10\frac{7}{8}$	3 0	3 $11\frac{1}{2}$	3 0	2 0	3 0	3	9	1 8	1 3	6 1	2	3	0.42	33.1
3 0	5 6	3 8	5 $4\frac{3}{4}$	4 0	5 $1\frac{7}{8}$	3 0	2 0	3 0	3	9	1 8	1 3	7 $3\frac{1}{2}$	2	3	0.61	50.4
4 0	6 0	4 0	5 $10\frac{5}{8}$	5 0	6 $4\frac{1}{4}$	3 0	2 0	3 0	3	9	2 0	1 6	8 $10\frac{3}{4}$	2	3	1.3	67.9
5 0	6 6	4 4	6 $4\frac{1}{2}$	6 0	7 $6\frac{5}{8}$	3 0	2 0	3 0	3	9	2 0	1 6	10 $1\frac{1}{4}$	2	3	1.6	85.6
6 0	7 0	4 8	6 $10\frac{3}{8}$	7 0	8 9	3 0	2 0	3 0	3	9	2 0	1 6	11 $3\frac{1}{2}$	2	3	2.6	103.5
7 0	7 6	5 0	7 $4\frac{1}{4}$	8 0	9 $11\frac{3}{8}$	3 0	2 0	3 0	3	9	2 0	1 6	12 6	2	3	3.0	121.4
8 0	8 0	5 4	7 $10\frac{1}{8}$	9 0	11 $1\frac{3}{4}$	3 0	2 0	3 0	3	9	2 0	1 6	13 $8\frac{1}{4}$	2	3	3.5	139.5

FIGURE 4-4 *Plan, elevation, and dimensions of the Parshall flume, from R. L. Parshall, "Measuring Water in Irrigation Channels with Parshall Flumes and Small Weirs," U.S. Soil Conservation Service, 1950. (Continued)*

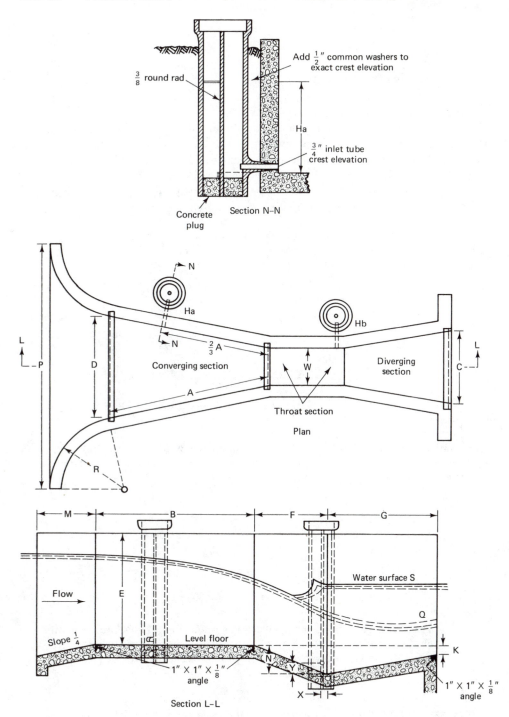

FIGURE 4-4   *(Continued)*

where Q = free discharge, cfs

W = throat width, ft

H_a = gage reading, ft

When the value of H_b/H_a exceeds 0.6 for 3-, 6-, and 9-in. flumes and exceeds 0.7 for 1- to 8-ft flumes, the flume is said to be submerged. When a flume is submerged, the actual discharge is less than that given by the above equations. The diagrams given in Figs. 4-5(a–c) can be used to determine discharge for submerged 3-, 6-, and 9-in. weirs. Figure 4-5(d) can be used to find a correction factor for flumes with 1-ft throat widths. The correction for the 1-ft flume can be made applicable to larger flumes by multiplying by the appropriate correction factors given below:

Size of flume (W, ft)	Correction factor
1	1.0
1.5	1.4
2	1.8
3	2.4
4	3.1
6	4.3
8	5.4

Use of the equations and correction diagrams is illustrated in the following example problems.

EXAMPLE PROBLEM 4-4:   Determine the discharge of a 6-in. Parshall flume for a gage depth H_a of 1.2 ft if (a) the flume has free discharge, and (b) if the flume is operating at a submergence of 0.8.

Solution:

1. The discharge for an unsubmerged 6-in. flume with an H_a of 1.2 ft is calculated from Eq. (4-5).

$$Q = 2.06\, H_a^{1.58}$$
$$= 2.06(1.2)^{1.58} = \textbf{2.75 cfs}$$

2. The discharge for 0.8 submergence is determined from Fig. 4-5(b). Enter the plot at a value of H_b/H_a of 0.80 and proceed horizontally to the line representing an H_a = 1.2. Drop vertically downward and read a discharge of

$$Q = \textbf{2.28 cfs}$$

EXAMPLE PROBLEM 4-5:   Determine the discharge through a 1-ft Parshall flume with gage reading H_a of 1.80 ft and a gage reading H_b of 1.48 ft.

Solution:

1. Calculate the ratio H_b/H_a to get the percentage of submergence.

$$H_b/H_a = \frac{1.48}{1.8} = 0.82 \text{ or } \textbf{82\%}$$

This value exceeds 0.7 so the flume is submerged.

2. Calculate the free discharge using Eq. (4-7).

$$Q = 4W\,H_a{}^{(1.522W)^{0.26}}$$
$$Q = 4(1)(1.8)^{(1.522(1))^{0.26}}$$
$$= 4(1.92) = \textbf{7.70 cfs}$$

3. Determine the appropriate correction factor using Fig. 4-5(d). Enter Fig. 4-5(d) with a value of $H_a = 1.8$ ft and proceed horizontally to the submergence curve of 82%. Drop vertically downward and read a correction of **1.22 cfs.**

4. Calculate the actual discharge.

$$Q = 7.70 - 1.22 = \textbf{6.48 cfs}$$

The engineer must properly set the elevation of the crest of a Parshall flume in order to achieve the correct relationship between the flume and adjacent units. The amount of head loss in a flume must be known in order to set the crest elevation. The head loss is a function of the throat width, the discharge, and the percentage of submergence, and can be determined from the diagrams given in Fig. 4-6.

Ideally, a flume should be set so that the degree of submergence does not exceed 0.6 for the 3-, 6-, and 9-in. flumes and 0.7 for the 1-ft, and larger, flumes. In all cases the degree of submergence should not exceed 95% since the flume will not measure correctly if the submergence is greater (Parshall, 1950).

EXAMPLE PROBLEM 4-6 [taken from Parshall (1950)]:   Design a Parshall flume to handle a flow of 20 cfs in a channel of moderate grade if the depth of flow in the channel is 2.5 ft. An elevation view of the flume is shown in Fig. 4-7.

Solution:

1. This discharge can be measured by several size flumes, but for the sake of economy the smallest practical size should be selected. Let us first evaluate a 4-ft flume with $H_b/H_a = 0.7$.
   (a) Solve Eq. (4-7) to find H_a

$$Q = 4WH_a^{(1.522W)^{0.26}}$$
$$20 = 4(4)\,H_a^{(1.522(4))^{0.26}} = 16\,H_a^{1.599}$$
$$H_a = \left(\frac{20}{16}\right)^{0.625} = \textbf{1.15 ft}$$

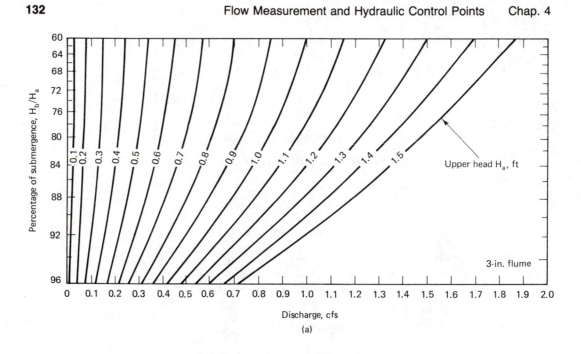

(a)

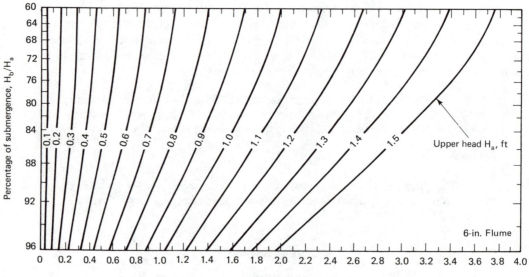

(b)

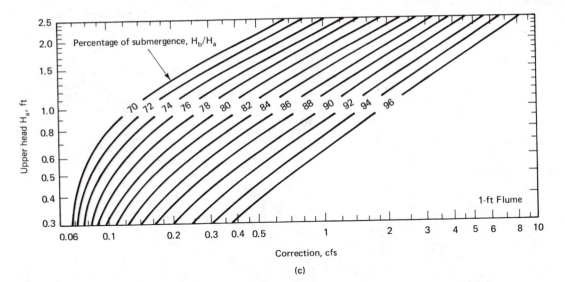

(c)

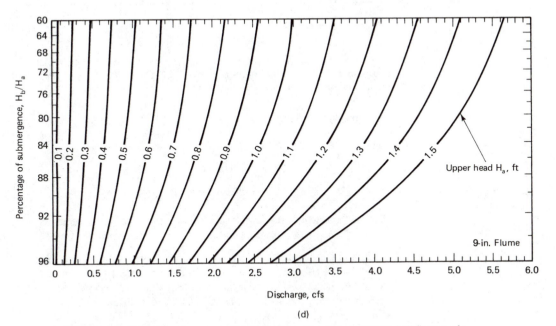

(d)

FIGURE 4-5 *Diagrams for computing flow through submerged Parshall flumes of various sizes, from* Open-Channel Hydraulics, *by Ven T. Chow, 1959. Used with permission of McGraw-Hill Book Company.*

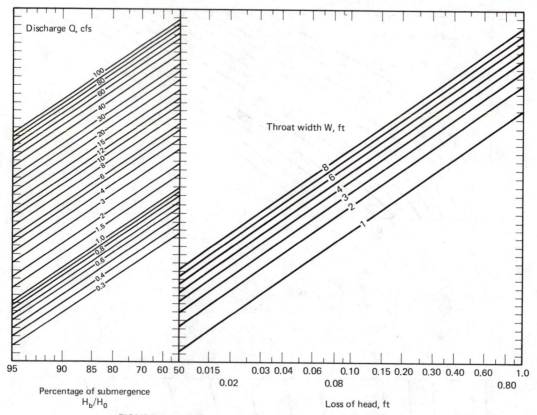

FIGURE 4-6 *Diagrams for determining the head loss through Parshall flumes of various sizes, from Open-Channel Hydraulics, by Ven T. Chow, 1959. Used with permission of McGraw-Hill Book Company. (Continued)*

(b)   If the submergence is to be set at 0.7

$$H_b/H_a = 0.7$$

$$H_b = H_a(0.7) = 1.15\,ft\,(0.7)$$

$$H_b = \mathbf{0.805} \approx \mathbf{0.81\ ft}$$

At 70% submergence the water surface in the throat, at H_b gage, is essentially level with the water surface at the lower end of the flume.

(c)   The depth in the channel downstream from the flume is 2.5 ft, and this depth will be regulated by some control point further downstream.

   If the H_b water surface is at the same elevation as the downstream water surface, the downstream channel must be set at distance, X, below the crest elevation.

$$X = 2.5 - 0.81 = \mathbf{1.69\ ft}$$

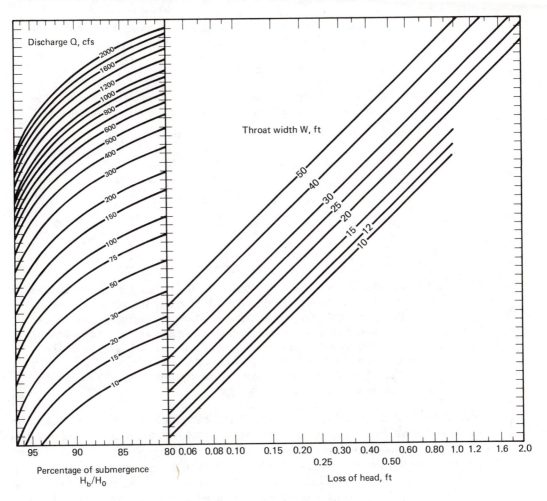

FIGURE 4-6   (Continued)

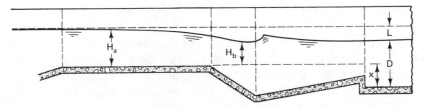

FIGURE 4-7   Elevation view of Parshall flume for Example Problem 4-6, from R. L. Parshall, "Measuring Water in Irrigation Channels with Parshall Flumes and Small Weirs," U.S. Soil Conservation Service, 1950.

(d)   The headloss through a 4-ft flume at a discharge of 20 cfs and a submergence of 70% can be determined from Fig. 4-6.

$$\text{Head loss} = \mathbf{0.40\ ft}$$

(e)   The water surface elevation at the upstream end of the flume can be determined by adding the head loss to the downstream water elevation.

2.   Similar calculations could be performed to evaluate a 2- and 3-ft flume.

	2-ft Flume (ft)	3-ft Flume (ft)
H_a	1.81	1.39
H_b	1.27	0.97
X	1.23	1.53
Head loss	0.62	0.48

When selecting the proper size flume, it will be necessary to evaluate the freeboard in the channel and the effect of the rise of the water surface upon the flow conditions. If it is found that these conditions are satisfactory, the 2-ft flume will be most economical because of its small size. However, when the width of the channel is considered, the final selection may favor the 3- or 4-ft flume, because moderate or long wing walls may be required for a small structure. Usually, the throat width of the flume will be from one-third to one-half of the channel width.

A Parshall flume offers certain advantages compared to weirs for measuring wastewater flows. For example, a flume can operate with partial submergence and thus does not require as much operating head as a freely discharging weir. In addition, the velocity through the flume is higher than the velocity in the approach channel, thereby greatly reducing the possibility of solids' deposition and permitting flow measurements in liquid streams containing suspended solids. Consequently, Parshall flumes are frequently selected as influent flow-measuring devices at wastewater treatment plants. When used in this capacity, they may also serve as the flow-control device for horizontal, gravity-type grit chambers.

4.3   VENTURI METERS

A Venturi meter, shown in Fig. 4-8, can be used to measure flows in pipes that flow under pressure. The device consists of an entrance cone, a short cylindrical section, and a diffuser cone that expands back to the full pipe diameter. As flow passes through the constricted throat of the meter, an increase in velocity and a decrease in pressure occur. The decrease in pressure is directly related to flow rate and, if measured, can serve as a basis for calculating the discharge. The equation for a Venturi meter, neglecting energy losses, is

$$Q = \frac{A_1 A_2 \sqrt{2g(H_1 - H_2)}}{\sqrt{A_1^2 - A_2^2}}$$ (4-8)

where

A_1 = area at upstream end of meter, ft^2

A_2 = area at throat of meter, ft^2

g = acceleration due to gravity, ft/s^2

$H = (H_1 - H_2)$ = pressure drop in meter, ft

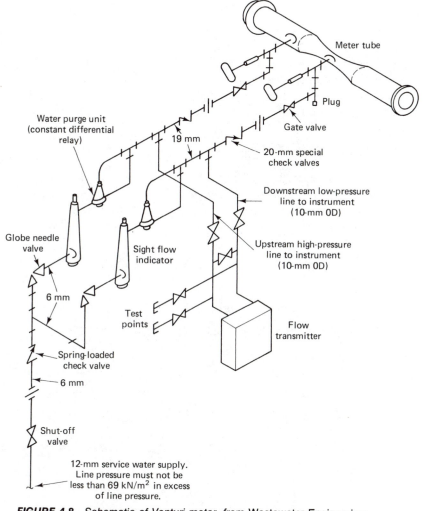

FIGURE 4-8   *Schematic of Venturi meter, from* Wastewater Engineering: Collection and Pumping of Wastewater, *by Metcalf and Eddy, Inc., 1981. Used with permission of McGraw-Hill Book Company.*

when friction losses are considered, Eq. (4-8) can be written as

$$Q = CA_2 \sqrt{2gH} \qquad\qquad \textbf{(4-9)}$$

where $C = C_1 C_2$

$$C_1 = \frac{A_1}{\sqrt{A_1^2 - A_2^2}}$$

C_2 = coefficient of energy loss

For standard meter tubes the value of C_1 ranges between 1.0062 and 1.0328 and the value of C_2 varies from 0.97 to 0.99. Thus, the range of values of C is from 0.98 to 1.02 (Metcalf and Eddy, 1981).

Venturi meters are frequently used as flow-measuring devices in water plants and also in wastewater plants, if a pressure line is available for their installation. When used to measure wastewater flows, manual cleaning rods or continuous-flushing systems must be provided to keep the meter clean and in good operating condition.

4.4 MAGNETIC FLOW METERS

A magnetic flow meter consists of a set of magnetic coils and an electrode assembly incorporated into a section of pipe as shown in Fig. 4-9. The device operates on the principle that when an electrical conductor passes through an electromagnetic field, an electromotive force is induced in the conductor that is proportional to the velocity of

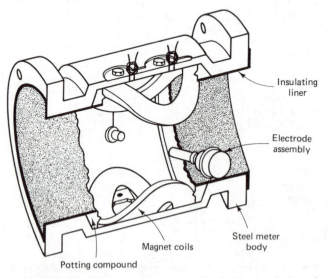

FIGURE 4-9 *Magnetic flow meter. Used with permission of Fischer and Porter.*

the conductor. In practice, magnetic coils generate the electromagnetic field and the liquid in the pipe (water or wastewater) serves as the conductor. The induced voltage is measured by the electrodes and is converted into a flow rate by appropriate electronic devices. If the pipe material is a conductor, the electrodes need not penetrate the pipe wall and the device is well suited to measuring the flow of solids-bearing liquids. Magnetic flow meters are thus used to measure the flow of raw wastewater and are frequently used to measure the flow of waste or return activated sludge. They are normally available in sizes varying from 2 to 24 in. and are reportedly accurate to ± 1% of the maximum scale reading for a velocity range of 3 to 30 fps.

4.5 HYDRAULIC CONTROL POINTS

A hydraulic control point can be defined as a point in a flow system at which there is a known relationship between depth and discharge. Several control points will exist in any water or wastewater treatment plant, and each will control the hydraulic profile through a portion of the plant.

According to the above definition, a control point could be provided by a weir or by the existence of critical depth, since the head on a weir or the value of critical depth can be calculated for any given flow. In fact, these are the two most common types of control points encountered in plant design. The concept of a control point is demonstrated in the following example problems.

EXAMPLE PROBLEM 4-7:   Water flows into a rectangular tank at a rate of 4 cfs and discharges from the tank over an adjustable rectangular weir. Determine the depth of water in the tank if the crest of the weir is located 6 ft above the bottom of the tank and the weir is 10 ft long.

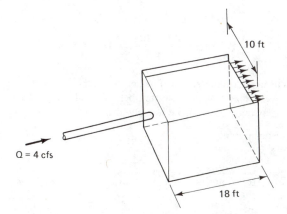

Solution: The head on a rectangular weir can be calculated from Eq. (4-1).

$$Q = 3.33L\, H^{3/2}$$

where $Q = 4.0$ cfs

L = length of the weir = 10 ft

H = the head on the weir, ft.

Rearranging and solving for H.

$$H = \left[\frac{Q}{3.33(L)}\right]^{2/3} = \left[\frac{4.0}{3.33(10)}\right]^{2/3}$$

$$H = \textbf{0.24 ft}$$

Thus the depth of water in the tank will be 6.243 ft. Although the head on the weir will remain constant as long as the flow rate does not change, the depth of water in the tank can be varied by raising or lowering the weir. Thus, the weir serves as the control point that determines the elevation of the water surface in the tank.

EXAMPLE PROBLEM 4-8:   A circular clarifier that is 30 ft in diameter receives an inflow of 500,000 gpd. The clarifier is equipped with a circular effluent weir that is composed of 90° V-notches located at 6 in. on center. The edge of the weir is placed 1.5 ft inside the wall of the clarifier so that the diameter of the weir is 27 ft.

At what elevation should the bottom of the V-notches be set in order to maintain the water surface in the tank at elevation 125.00 ft.

Solution:

1.   Determine the length of the weir and the number of V-notches in the weir.

$$\text{Weir length} = \pi(\text{diameter}) = \pi(27 \text{ ft})$$

$$= 84.78 \text{ ft.} \simeq 84.5 \text{ ft}$$

$$\text{Number of notches} = 84.5 \text{ ft} \times \frac{2 \text{ notches}}{\text{ft}} = \textbf{169 notches}$$

Note: A scum box will normally be attached to the weir and will occupy a portion of the weir thus reducing the number of available notches. The presence of a scum box will be neglected in this example.

2.   Calculate the discharge per notch.

$$\text{Discharge per notch} = \frac{500,000 \text{ gpd}}{169 \text{ notches}} = 2959 \frac{\text{gpd}}{\text{notch}}$$

$$= 2959 \frac{\text{gpd}}{\text{notch}} \times \frac{1 \text{ ft}^3}{7.48 \text{ gal}} \times \frac{1 \text{ day}}{1400 \text{ min}} \times \frac{1 \text{ min}}{60 \text{ sec}}$$

$$= \textbf{0.0046} \frac{\textbf{cfs}}{\textbf{notch}}$$

3. Calculate the head on each notch using Eq. (4-3) V-notch weirs.

$$Q = 2.5\,H^{2.5}$$

$$H = \left[\frac{Q}{2.5}\right]^{1/2.5} = \left[\frac{0.0046}{2.5}\right]^{0.4}$$

$$H = \textbf{0.08 ft}$$

4. Calculate the elevation of the bottom of the V-notches that will place the water surface in the clarifier at elevation 125.00.

$$\text{V-notch elevation} = 125.00 - 0.08 = \textbf{124.92 ft.}$$

4.6 PLANT HYDRAULICS

According to the fundamentals of hydraulics, subcritical flow is controlled by downstream control points, and supercritical flow is controlled by upstream control points. Downstream conditions cannot influence supercritical flow, because downstream disturbances cannot be propogated back upstream against supercritical velocities. Water and wastewater treatment plants normally involve subcritical flow conditions; thus hydraulic profiles are primarily determined by downstream control points.

A typical profile through a wastewater treatment plant is shown in Fig. 4-10, and the various control points are identified. For example, the effluent weir in the primary clarifier serves as one control point and controls the hydraulic profile in the clarifier and back through the effluent channel of the aerated grit basin. Once the flow passes over the weir in the primary clarifier it is controlled by the next downstream control point, the influent weir in the aeration basins.

When designing a wastewater treatment plant the engineer must calculate the head losses through the various units in the plant and must set the elevations of the control points to produce the calculated hydraulic profile. The starting point for the design depends upon the topography of the plant site. If abundant head is available, i.e., the plant is located at an elevation well above the receiving body of water, the engineer may start calculating losses at the headworks of the plant and then set the control points at the proper elevations as he proceeds down through the plant. If the plant site is such that excess head is not available, the design may start from the high water level in the receiving body of water and proceed, in the reverse direction, through the plant. Head losses are calculated, and control point elevations are set as before.

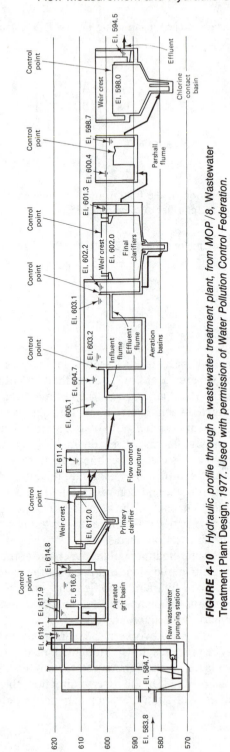

FIGURE 4-10 *Hydraulic profile through a wastewater treatment plant, from MOP/8, Wastewater Treatment Plant Design, 1977. Used with permission of Water Pollution Control Federation.*

REFERENCES

METCALF and EDDY, INC. *Wastewater Engineering: Collection and Pumping of Wastewater,* McGraw–Hill Book Company, New York, New York (1981).

PARSHALL, R. L., "Measuring Water in Irrigation Channels with Parshall Flumes and Small Weirs," U. S. Soil Conservation Service, Circular 843, May 1950.

PARSHALL, R. L., "Parshall Flumes of Large Sizes," Colorado Agricultural Experimentation, Bulletin No. 386, May 1932; revised as Bulletin No. 426A, March, 1953.

VENNARD, J. K., *"Elementary Fluid Mechanics,"* 4th Edition, John Wiley & Sons, Inc., New York, New York (1966).

VENNARD, J. K., and WESTON, R. F., "Submerged Effect on Sharp-Crested Weirs," *Engineering News Record,* p. 118, June 3, 1943.

WATER POLLUTION CONTROL FEDERATION, *Wastewater Treatment Plant Design,* Washington, D. C. (1977).

5

PUMPS

5.1 BASIC CONCEPTS

This chapter will discuss the general characteristics of pumps and the design of pumping systems for water and wastewater treatment facilities. Only centrifugal pumps will be considered, since they are the principal type used in treatment plants. Detailed discussions of screw pumps and positive displacement pumps are given by Karassik et al. (1976) and Tchobanoglous (1981). Transporting water or wastewater from one location to another requires pumping when gravity flow alone will not deliver the desired flow rate or capacity. Typically, in fact, it is necessary to transport liquids from a lower to a higher elevation, and a pump is needed if any flow is to be realized. The pump selected for a specific installation must deliver sufficient energy to the liquid at the specified rate of flow to overcome the change in elevation head and the losses in the piping system.

A definitive sketch of a simple pump installation is shown in Fig. 5-1. The average velocity in the inlet and outlet pipes to the pump are called the suction velocity, V_s, and the discharge velocity, V_D, respectively. The one-dimensional energy equation can be written between points 1 and 4 for the situation shown in Figure 5-1 as follows

$$\frac{V_1^2}{2g} + Z_1 + \frac{P_1}{\gamma} + h_p = \frac{V_4^2}{2g} + Z_4 + \frac{P_4}{\gamma}$$
$$+ \Sigma(\text{minor losses}) + \Sigma(\text{friction losses})$$

(5-1)

where h_p is the net head delivered by the pump to the liquid. This term is called the pump head. The loss terms were discussed in detail in Chap. 1. By assuming that the velocity head terms are negligible at points 1 and 4 in the two tanks, Eq. (5-1) can be simplified and rearranged to give

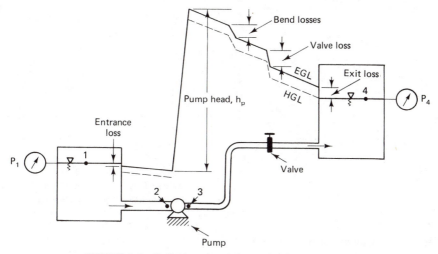

FIGURE 5-1   *Definitive sketch for a simple pump installation.*

$$h_p = \left[\frac{p_4}{\gamma} - \frac{p_1}{\gamma}\right] + [Z_4 - Z_1] + \Sigma(\text{minor losses}) + \Sigma(\text{friction losses}) \quad \textbf{(5-2)}$$

or $h_p = [\text{pressure head}] + [\text{elevation head}] + [\text{total head loss}]$ **(5-3)**

or $h_p = [\text{static head}] + [\text{total head loss}]$ **(5-3a)**

The sum of the static head and the total head loss is called the system head.

System-Head Curve

Since friction and minor loss terms which comprise the total head loss of a system vary with capacity, Q, the system head of a pipeline is also a function of capacity. A curve showing the relationship between the system head and capacity (Fig. 5-2) is called the *system-head curve*. This type of curve can be constructed for a specific system using the methods presented in Chap. 1 to calculate the minor and friction loss terms for several values of capacity and then plotting h_p versus Q using Eq. (5-2).

Pump-Head Curve

A centrifugal pump operating at a constant rotative speed, has a unique relationship between the head developed by the pump and the capacity. A plot of this relationship is called a *pump curve*. Figure 5-3 shows a pump curve and a system-head curve for a pumping system. The intersection of the two curves represents the actual operating conditions of the system.

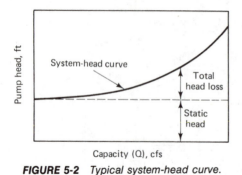

FIGURE 5-2 Typical system-head curve.

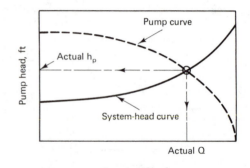

FIGURE 5-3 System-head curve and pump curve for typical pump installation.

Pump Characteristics Curves

The operating point, indicated by the intersection of the two curves in Fig. 5-3 may not correspond to the most efficient operating condition for the pump. Therefore, it is necessary to examine the efficiency and power requirements for the pump being considered also. A group of curves showing head–capacity, efficiency–capacity, and power input–capacity relationships for a specific pump are called *pump characteristics curves* (Fig. 5-4). These curves are developed during pumping tests performed by pump manufacturers and are available on request. The horsepower delivered by a pump to the liquid is given by the equation

$$HP = \frac{\gamma Q h_p}{550} \qquad (5\text{-}4)$$

where γ = sp. wt. of liquid in lb/ft^3, Q = discharge in cfs, and h_p = pump head in feet. The horsepower that must be delivered to a pump is obtained by dividing Eq. (5-4) by the pump efficiency as follows:

$$HP = \frac{\gamma Q h_p}{550 \times \text{efficiency}} \qquad (5\text{-}5)$$

Pump characteristics curves should be used to choose a pump that will deliver the desired flow rate for a system at an efficient and economical level of operation for the type of pump being considered.

Figure 5-5 shows a series of pump curves corresponding to different rotative speeds for a particular centrifugal pump as determined from laboratory tests. Contours of equal efficiency are also shown on the figure. It is, of course, desirable to choose a pump that will operate near its optimum efficiency. Figures similar to Fig. 5-5 relating head, discharge, and efficiency for different impeller diameters (rather than rotative speeds) are also commonly used in pump selection.

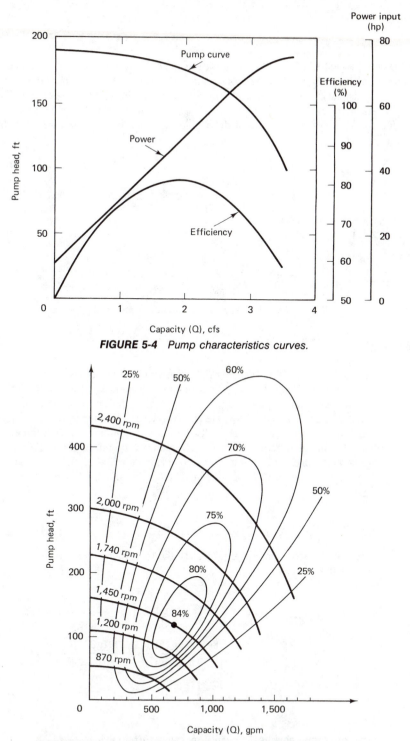

FIGURE 5-4 *Pump characteristics curves.*

FIGURE 5-5 *Characteristics of a centrifugal pump at various speeds of rotation with contours of equal efficiency, from* Fluid Mechanics with Engineering Applications, seventh edition, *by Daugherty and Franzini, 1977. Used with permission of McGraw-Hill Book Company.*

EXAMPLE PROBLEM 5-1: Water at 60°F is to be delivered from tank A to tank B as shown in Fig. 5-6 at a flow rate of about 3 cfs. The 150-ft-long suction and the 400-ft-long discharge pipes are 8- and 6-in. diameter cast-iron pipe, respectively. The pump characteristics curves for a pump being considered are shown in Fig. 5-4. Develop the system-head curve and determine the suitability of the pump for this system.

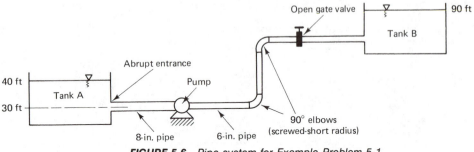

FIGURE 5-6 *Pipe system for Example Problem 5-1.*

Solution:

1. Develop a relationship between flow rate and friction losses.

    Since the friction factor, f, is a function of both the Reynold's number, Re, and the relative pipe roughness, e/D, the friction factor must be estimated initially, since the flow rate is unknown and Re cannot be determined. This is done by assuming that the flow is completely turbulent and, therefore, that f is only a function of e/D. The validity of this assumption can be checked later and adjustments in f can be made then if necessary. The relative roughness for cast iron is 0.0102 in. (Table 1-1, p. 12). For the 8-in. pipe $e/D = \dfrac{0.0102}{8} = 0.0013$ and, from Fig. 1-7, $f = 0.021$. For the 6-in. pipe $e/D = \dfrac{0.0102}{6} = 0.0017$ and, from Fig. 1-7, $f = 0.022$. Using the Darcy–Weisbach equation [Eq. (1-20)], the friction losses in the pipes can be expressed as follows:

$$\Sigma(\text{Friction losses}) = \left[\frac{8f_8 L_8}{\pi^2 g D_8^5} + \frac{8f_6 L_6}{\pi^2 g D_6^5}\right] Q^2$$

$$= \left[\frac{8(0.021)(150)}{\pi^2 (32.2)(0.667)^5} + \frac{8(0.022)(400)}{\pi^2 (32.2)(0.5)^5}\right] Q^2$$

$$= \mathbf{7.69Q^2}$$

where Q is in cfs.

2. Develop a relationship between flow rate and minor losses.

    Table 1-7 is used to determine K values for the appurtenances in the pipe system. The minor losses can be expressed as

$$\Sigma(\text{Minor losses}) = \left[\frac{8}{\pi^2 g D_6^4}(K_{\text{ent}}) + \frac{8}{\pi^2 g D_6^4}(2K_{\text{bend}} + K_{\text{valve}} + K_{\text{exit}})\right]Q^2$$

$$= \left[\frac{8}{\pi^2(32.2)(0.667)^4}(0.5) + \frac{8}{\pi^2(32.2)(0.5)^4}(1.8 + 0.19 + 1.0)\right]Q^2$$

$$= \mathbf{1.27 Q^2}$$

where Q is in cfs.

3. Estimate the operating head and capacity.

    Equation (5-3) is used to determine the system-head relationship as follows, where the pressure head is zero

$$h_p = \text{static head} + \text{total head loss}$$

$$= (90 - 40) + 7.69 Q^2 + 1.27 Q^2$$

$$h_p = \mathbf{50 + 8.96 Q^2}$$

where h_p is in feet and Q is in cfs.

This equation is plotted on the pump characteristics curves as shown in Fig. 5-7. The operating head and capacity, corresponding to the intersection of the system-head curve and the pump curve are 139 ft and 3.13 cfs, respectively. Although the pump will deliver the desired flow rate, the

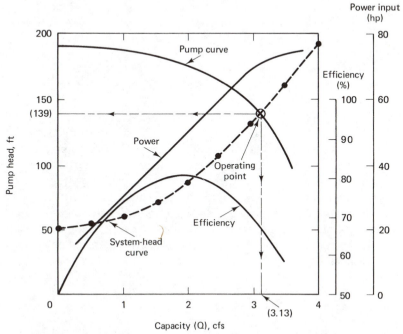

FIGURE 5-7 *System curve for Example Problem 5-1 plotted on pump characteristics curves.*

efficiency for the operating condition, 67%, is well below the maximum operating efficiency of the pump. Consequently, the pump considered should not be chosen for this system.

4. Check the validity of the initial f value.

The f values can be checked at this time for $Q = 3.13$ cfs. From Table 1-3 the viscosity of water at 60°F is 1.217×10^{-5} ft^2/s. Hence, for the 8-in. pipe

$$\text{Re}_8 = \frac{4Q}{\pi D_8 \nu} = \frac{4(3.13)}{\pi(0.667)(1.217 \times 10^{-5})} = \mathbf{4.9 \times 10^5}$$

The corrected value of the friction factor of the 8-in. pipe ($e/D = 0.0013$) is read from Fig. 1-7 as about 0.021. Following a similar procedure for the 6-in. pipe yields an f value of 0.022. Consequently, the f values for both pipes are essentially unchanged, and the pump evaluation described above is valid. If the f values had changed significantly the entire evaluation should be repeated using the new f values.

Pump Similarity

Dimensional analysis can be used to develop equations that relate the performance of geometrically similar pumps. These equations, called *affinity laws,* are quite useful in predicting the performance of prototype pumps using the results from pumping tests on scaled-down pumps. The affinity laws will not be derived here, but are given as follows:

$$h_p = \zeta_{h_p} D^2 n^2 \tag{5-6}$$

$$Q = \zeta_Q D^3 n \tag{5-7}$$

$$P = \zeta_p D^5 n^3 \tag{5-8}$$

where h_p = pump head, Q = capacity, P = power, D = impeller diameter, n = rotative speed in rpm, and ζ_{h_p}, ζ_Q, and ζ_P are performance coefficients. Consequently, by assuming that efficiency is independent of pump size, the affinity laws for a particular pump design can be determined from test data on a small-scale pump and can then be used to construct pump curves for prototype pumps.

The affinity laws can also be used to determine pump curves for a specific pump at different speeds. For instance, if the pump head and power curves are known for a centrifugal pump at rotative speed, n_1, the following relationships, derived from Eq. (5-6), (5-7), and (5-8), can be used to construct the pump and power curves for another rotative speed, n_2,

$$Q_2 = Q_1(n_2/n_1) \tag{5-9}$$

$$h_{p_2} = h_{p_1}(n_2/n_1)^2 \tag{5-10}$$

$$P_2 = P_1(n_2/n_1)^3 \tag{5-11}$$

TABLE 5-1   *Given and calculated pump curve data for Example Problem 5-2.*

n = 900 rpm		n = 700 rpm	
Capacity (gpm)	*Head (ft)*	*Capacity (gpm)*	*Head (ft)*
(1)	*(2)*	*(3)*	*(4)*
0	62.5	0	37.8
400	62.7	311	37.9
800	62.5	633	37.8
1,200	61.2	934	37.0
1,600	57.7	1,244	34.9
2,000	52.4	1,556	31.7
2,400	43.6	1,867	26.4
2,800	30.8	2,178	18.6

EXAMPLE PROBLEM 5-2:   Columns 1 and 2 of Table 5-1 give the pump curve data for a centrifugal pump operating at 900 rpm. The pump curve is also shown graphically in Fig. 5-8. Determine the pump curve for the pump operating at a rotative speed of 700 rpm.

Solution:

1. Develop similarity equations.
   For this example, Eq. (5-9) and (5-10) become

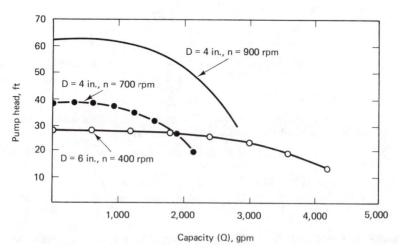

FIGURE 5-8   *Given and calculated pump curves for Example Problems 5-2 and 5-3.*

$$Q_2 = \left[\left(\frac{n_2}{n_1} \right) Q_1 \right] = \left[\left(\frac{700}{900} \right) Q_1 \right] = [0.778Q_1]$$

and

$$h_{p_2} = \left[\left(\frac{n_2}{n_1} \right)^2 h_{p_1} \right] = [(0.778)^2 h_{p_1}] = [0.605 h_{p_1}]$$

where the subscripts 1 and 2 refer to operating conditions at $n = 900$ rpm and $n = 700$ rpm, respectively.

2. Use similarity equations to determine pump curve for $n = 700$ rpm. The relationships are used to calculate the values presented in Columns 3 and 4 of Table 5-1 and represented by the short curve in Fig. 5-8.

EXAMPLE PROBLEM 5-3: If the pump in Example Problem 5-2 has a 4-in. impeller, determine the pump curve for a geometrically similar pump with a 6-in. impeller operating at a rotative speed of 400 rpm.

Solution:

1. Develop similarity equations. Operational points for the pump curve of the larger pump can be determined from the original pump curve using the following relationships

$$h_{p_6} = \left[\left(\frac{D_6 n_6}{D_4 n_4} \right) \right]^2 h_{p_4} = \left[\frac{\left(\frac{6}{12} \right) 400}{\left(\frac{4}{12} \right) 900} \right]^2 h_{p_4} = (0.444 h_{p_4}) \quad (5\text{-}12)$$

and

$$Q_6 = \left[\frac{D_6^3 n_6}{D_4^3 n_4} \right] Q_4 = \left[\frac{\left(\frac{6}{12} \right)^3 400}{\left(\frac{4}{12} \right)^3 900} \right] Q_4 = (1.50 Q_4) \quad (5\text{-}13)$$

which were derived from Eqs. (5-6) and (5-7). The subscripts 6 and 4 refer to the larger and the smaller pumps, respectively.

2. Use similarity equations to determine pump curve for $n = 400$ rpm and $D = 6$ in.
    Columns 3 and 4 of Table 5-2 present the pump curve for the larger pump in tabular form as determined from the values in Columns 1 and 2 using Eqs. (5-12) and (5-13).

EXAMPLE PROBLEM 5-4: The system curve for a system in which the pump described in Example Problem 5-2 is used is given below in tabular form. Determine the pump capacity and efficiency for $n = 900$ rpm and $n = 700$ rpm.

TABLE 5-2 *Given and calculated pump curve data for Example Problem 5-3.*

n = 900 rpm, D = 4 in.		n = 400 rpm, D = 6 in.	
Capacity (gpm)	*Head (ft)*	*Capacity (gpm)*	*Head (ft)*
(1)	*(2)*	*(3)*	*(4)*
0	62.5	0	27.7
400	62.7	600	27.8
800	62.5	1,200	27.7
1,200	61.2	1,800	27.2
1,600	57.7	2,400	25.6
2,000	52.4	3,000	23.3
2,400	43.6	3,600	19.4
2,800	30.8	4,200	13.7

Discharge (cfs)	0	400	800	1,200	1,600	2,000	2,400	2,800
Head (ft)	15	16.5	20.9	28.3	36.7	52.0	68.3	87.5

Solution:

1. Plot system curve, and pump curves for $n = 700$ and 900 rpm. The pump curves for the two rotative speeds are shown together with the system curve corresponding to the data given above in Fig. 5-9. The efficiency of the pump for $n = 900$ rpm is also shown graphically in the figure.

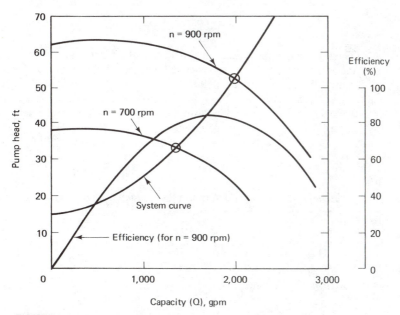

FIGURE 5-9 *Pump curves, system curve, and efficiency curve for n = 900 rpm (Example Problem 5-4).*

2. Determine operating head conditions and efficiency for $n = 900$ rpm. The operating condition for $n = 900$ rpm is read from Fig. 5-9 as $Q = 2010$ gpm at a head of 51.6 ft. The efficiency at this operating condition is read directly from the curve as 81%.

3. Determine operating conditions and efficiency for $n = 700$ rpm. The operating conditions for $n = 700$ rpm are $= 1370$ gpm at a head of 33.0 ft. In order to estimate the pump efficiency for these conditions, it is necessary to determine the corresponding point on the pump curve for $n = 900$ rpm since the efficiency is known for the higher rotative speed. By combining Eqs. (5-9) and (5-10) the rotative speed ratio, n_2/n_1, can be eliminated to give the relationship

$$h_{p_2} = h_{p_1}\left(\frac{Q_2}{Q_1}\right)^2$$

or, letting $h_{p_2} = h_p$ and $Q_2 = Q$

$$h_p = \left[\frac{h_{p_1}}{Q_1^2}\right]Q^2$$

This equation is a parabola passing through the origin and through (h_{p_1}, Q_1) which describes the locus of corresponding points for the pump at different rotative speeds. That is, each of the points on the parabola corresponds to operating conditions having the same pump efficiency. The parabola passing through the operating point for $n = 700$ gpm is, therefore,

$$h_p = \left[\frac{33.0}{(1,370)^2}\right]Q^2 = \frac{Q^2}{56,876}$$

This equation is plotted in Fig. 5-10. It passes through the operating point of the pump curve for $n = 700$ rpm and intersects the pump curve for $n = 900$ rpm at the corresponding point of $Q = 1790$ gpm and $h_p = 55.1$ ft. From Fig. 5-9, the efficiency for $Q = 1790$ gpm at $n = 900$ rpm is read as 82%.

All of the points on the dashed line in Fig. 5-10 have the same specific speed, N_s, defined by the equation

$$N_s = \frac{nQ^{0.5}}{h_p^{0.75}}$$

where $n =$ rotative speed in rpm, $Q =$ capacity in gpm, and $h_p =$ head in ft. For example, at $n = 700$ rpm

$$N_s = \frac{700(1370)^{0.5}}{(33)^{0.75}} = 1880$$

and at $n = 900$ rpm

$$N_s = \frac{900(1790)^{0.5}}{(55.1)^{0.75}} = 1880$$

The concept of specific speed will be discussed further in the next section.

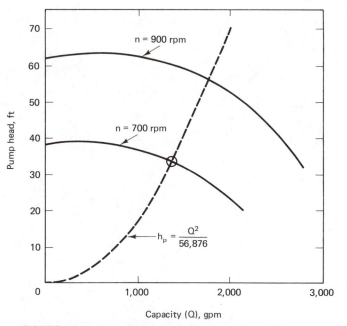

FIGURE 5-10 *Pump curves and line of constant efficiency passing through operation point at n = 700 rpm.*

5.2 *TYPES OF CENTRIFUGAL PUMPS*

As water passes through a centrifugal pump, it is accelerated by a rotating series of vanes, called an impeller. After leaving the impeller, the water is decelerated in a series of stationary guide vanes or in an expanding section of the pump casing called a volute, before it enters the discharge line. During the deceleration phase, most of the additional velocity head acquired in the impeller is converted to pressure head in accordance with Bernoulli's equation. Consequently, there is a net increase in pressure head from the inlet to the outlet of a centrifugal pump.

The most common classification of centrifugal pumps is based on the major direction of outflow from the pump impeller relative to the axis of rotation. The pump types are radial, axial (or propeller), and mixed flow. Most applications in water and wastewater treatment call for radial- and mixed-flow pumps although, axial-flow pumps are well suited for transporting relatively clean water at large flow rates subject to low head conditions. The general application ranges for the different types of centrifugal pumps are indicated in Fig. 5-11.

Water enters the impeller of a radial-flow pump axially. In passing through the impeller, the water acquires tangential and radial velocity components relative to the axis of rotation and, theoretically, upon leaving the impeller, has no axial velocity component. Figure 5-12 shows a section of a radial-flow centrifugal pump with a

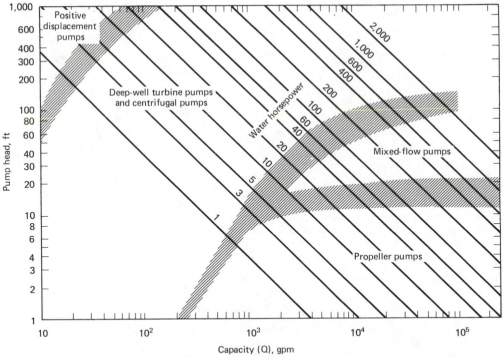

FIGURE 5-11   *Chart for selection of type of pump. Used with permission of Fairbanks, Morse and Company.*

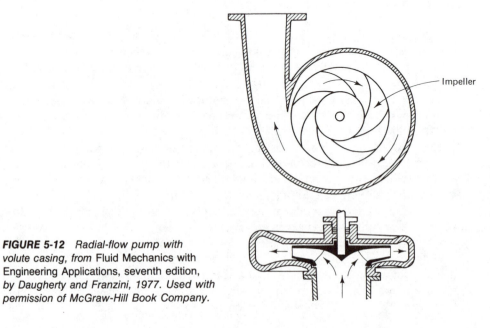

FIGURE 5-12   *Radial-flow pump with volute casing, from* Fluid Mechanics with Engineering Applications, *seventh edition, by Daugherty and Franzini, 1977. Used with permission of McGraw-Hill Book Company.*

volute casing. This type of pump equipped with a nonclog impeller is usually used to pump untreated wastewater. Nonclog impellers typically have only two or three vanes and are designed to pass up to 3-in. diameter solids. Figure 5-13 shows a radial-flow pump equipped with a series of stationary guide vanes, called a diffuser, rather than a volute casing, to decelerate the flow. This type of pump is frequently called a turbine pump.

Both the inflow and the outflow from an axial-flow pump are in the axial direction. The vanes of the impeller are designed so that the water is only accelerated tangentially in passing through the impeller. The additional kinetic energy of the water leaving an axial-flow impeller is, therefore, due to a large tangential velocity component. A series of stationary vanes, positioned just downstream from the impeller, acts to remove the "swirl" from the outflow and, thereby, convert the kinetic energy due to the tangential velocity component into pressure head. Figure 5-14 shows an axial-flow pump. Guide vanes are also usually positioned upstream from the impeller.

The flow through the impeller of a mixed-flow pump acquires both tangential and radial velocity components. Like the radial-flow pump, either a volute casing or a

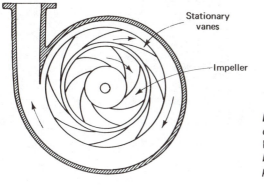

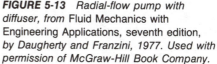

FIGURE 5-13 *Radial-flow pump with diffuser, from* Fluid Mechanics with Engineering Applications, *seventh edition, by Daugherty and Franzini, 1977. Used with permission of McGraw-Hill Book Company.*

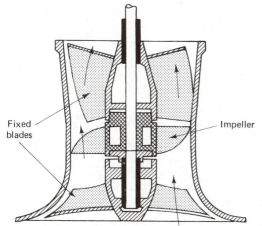

FIGURE 5-14 *Axial-flow pump. Used with permission of Ingersoll-Rand Company.*

diffuser is used to decelerate the flow and, thus, to transform kinetic energy into pressure head downstream from the impeller.

The concept of specific speed is useful in pump classification. The specific speed of a pump, N_s, is normally defined by the following equation:

$$N_s = \frac{nQ^{1/2}}{h_p^{3/4}} \qquad (5\text{-}14)$$

in which n = rotative speed in rpm, Q = discharge in gpm, and h_p = head produced by the pump in feet. Specific speed varies with impeller type as shown in Fig. 5-15. In identifying a pump, the specific speed corresponding to the operating conditions at maximum efficiency, called the type-specific speed, is used.

Pump curve shape varies with specific speed as shown in Fig. 5-16 where typical pump curves are shown for radial-, mixed-, and axial-flow pumps. It is interesting to note that at "shut-off" conditions, the power input is minimum for radial-flow pumps and maximum for axial-flow pumps. The steepness of power input curves for axial-

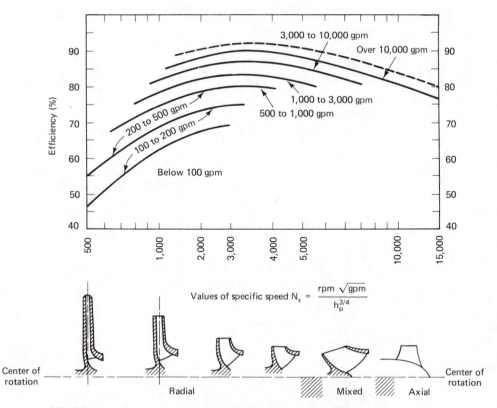

FIGURE 5-15 *Approximate relative impeller shapes and efficiency variations with specific speed, after Karassik and Carter (1960).*

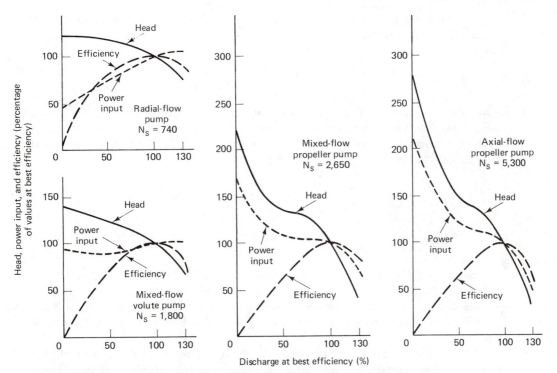

FIGURE 5-16 *Typical pump characteristics curves for centrifugal pumps, from* Wastewater Engineering: Collection and Pumping of Wastewater, *by Metcalf and Eddy, Inc., 1981. Used with permission of McGraw-Hill Book Company.*

flow pumps indicates that they should only be operated at conditions very near the optimum efficiency point. Throttling an axial-flow pump to deliver flow rates significantly less than the design flow can overload and damage the motor. Consequently, radial- and mixed-flow volute pumps are normally chosen when operation is required for a wide range of conditions.

5-3 CAVITATION

One of the principal problems encountered in pump applications is the development of cavitation in a pump. Cavitation occurs in the flow of a liquid when the pressure falls below the vapor pressure of the liquid and vapor "pockets" are formed. The occurrence of this phenomenon in a pump can reduce capacity and efficiency by constricting the inflow section at the pump impeller. Cavitation can also damage a pump by pitting the casing and impeller or by inducing excessive vibrations.

The term *net positive suction head* (NPSH) is useful in designing and evaluating pumping systems with regard to cavitation avoidance. NPSH is defined as the total head at the suction side of a pump (relative to a datum at the elevation of the pump inlet) minus the vapor pressure head of the liquid being transported. Mathematically, this definition is expressed as follows:

$$\text{NPSH} = \frac{P_s}{\gamma} + \frac{V_s^2}{2g} - \frac{P_v}{\gamma} \tag{5-15}$$

where P_s = absolute pressure at pump inlet, V_s = average velocity at pump inlet, and P_v = absolute vapor pressure of the liquid. The net positive suction head is more frequently defined in terms of the total head at Sec. 1 of Fig. 5-1 and the lift and losses from Sec. 1 to the pump inlet as follows:

$$\text{NPSH} = \frac{P_1}{\gamma} + \frac{V_1^2}{2g} - Z_s - \Sigma(\text{minor losses from 1 to 2})$$

$$- \Sigma(\text{friction losses from 1 to 2}) - \frac{P_v}{\gamma} \tag{5-16}$$

where Z_s = distance pump intake is above water surface in supply reservoir. Usually the velocity head term, $V_1^2/2g$, is negligible. It is important to remember that the pressure P_1, in Eq. (5-16) must also be expressed as absolute pressure. The value of NPSH determined from Eqs. (5-15) or (5-16) for a system is called the *available NPSH*.

A centrifugal pump will operate without cavitation provided the available NPSH of the system in which it is installed exceeds the *required NPSH* for the specific pump. The value of the required NPSH for a particular pump design is based on performance tests performed by the pump manufacturer. The vapor pressure head of water in absolute notation is given in Table 1-3 as a function of temperature.

EXAMPLE PROBLEM 5-5:   The pumping system shown in Fig. 5-17 is to deliver 4 cfs of water at 60°F. The required NPSH of the pump is 9 ft. The suction piping is constructed of 8-in. cast-iron pipe. Determine if cavitation will be a problem for the proposed system when P_{ATM} = 14.7 psia.

Solution:

1.   Determine the coefficients for minor and friction losses.
       The minor loss coefficients as given in Table 1-7 are 0.5 and 0.3, respectively, for the abrupt entrance and for each elbow. The friction factor is obtained from Fig. 1-7 for

$$e/D = \frac{(0.0102/12)}{(8/12)} = 0.00128$$

and $$\text{Re} = \frac{VD}{\nu} = \frac{4Q}{\pi D \nu} = \frac{4(4)}{\pi(0.667)(1.217 \times 10^{-5})} = \textbf{628,000}$$

as f = 0.021. The values for e and ν are given in Tables 1-1 and 1-3, respectively.

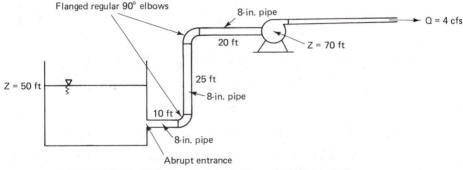

FIGURE 5-17   *Definition sketch for Example Problem 5-5.*

2. Determine the available NPSH.

The velocity in the 8-in. suction pipe is $V = Q/A = 4/\pi(0.333)^2 =$ 11.48 fps, and the vapor pressure from Table 1-3 is 0.59 ft. The available NPSH can be obtained from Eq. (5-16) as follows:

$$NPSH = \frac{(14.7)(144)}{62.4} + \frac{V_1^2}{2g} - (70 - 50) - (0.5 + 0.3 + 0.3)\frac{(11.48)^2}{2g}$$

$$- \frac{(0.021)(55)}{(0.667)}\frac{(11.48)^2}{2g} - 0.59 = \textbf{7.5 ft}$$

3. Compare available NPSH and required NPSH.

Since the available NPSH of the proposed system is less than the required NPSH of the pump, cavitation problems would be likely to occur. Consequently, either another pump should be chosen or the suction piping system should be altered to increase the available NPSH to at least 9 ft, and preferably a bit more.

5.4   MULTIPLE-PUMP OPERATION

Pumping systems are often required to perform at different discharge and total head conditions during the expected life of the pumps. These requirements can be met at acceptable levels of efficiency by operating several pumps in parallel or in series or, in some cases, by choosing multiple- or variable-speed motors.

When two pumps are operated in parallel, the head–capacity curve for the two-pump setup is obtained by adding the abscissas of the individual pump curves. This is illustrated in Fig. 5-18 where two different pumps are arranged in a parallel configuration. Including more pumps in the parallel system would further increase the total discharge passing through the loop for a given pressure head. It is sometimes recommended that nearly identical pumps be used in parallel systems to avoid problems, such as reverse flow, at low-capacity operation. "Mismatched" pumps will perform satisfactorily in parallel, however, if the shapes of the pump curves are similar and if the shut-off heads are nearly the same for each of the pumps and if the heads at the best efficiency points (bep), are about equal (Karassik and Carter, 1960).

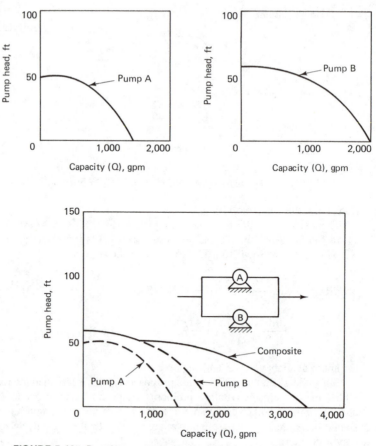

FIGURE 5-18 *Development of pump curves for two pumps operated in parallel.*

Operating pumps in series will increase the total head while approximately maintaining the same discharge as that for a single pump. The pump curve for a series of pumps is, therefore, obtained by adding the heads of all of the pumps for the same discharge. For example, if pumps A and B of Fig. 5-18 were operated in series, the pump curve for the two pumps would be as shown in Fig. 5-19. The performance concept of series-pump operation is the basis of multistage pumps in which several identical impellers are arranged in series within a single casing. A multistage pump produces the same capacity as a single impeller while developing a total head approximately proportional to the number of impellers, or stages. Multistage pumps are particularly useful in producing water at relatively high heads from wells where impeller diameter is limited by the well bore. A schematic of a five-stage vertical pump installed in a water well is shown in Fig. 5-20. Vertical multistage pumps are also desirable in industrial or treatment applications when floor space is limited.

Multiple-speed pumps are sometimes used when a system must deliver flow at different capacities. A multiple-speed pump is preferable to a parallel-pumping ar-

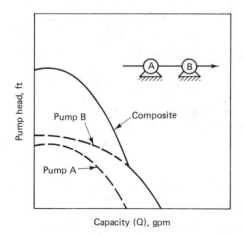
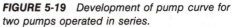

FIGURE 5-19 *Development of pump curve for two pumps operated in series.*

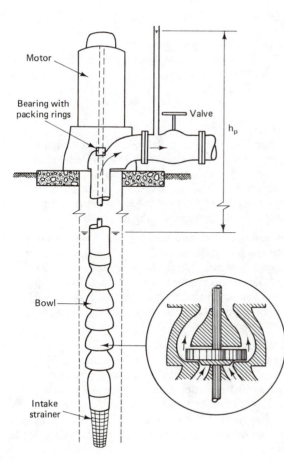

FIGURE 5-20 *Schematic of five-stage centrifugal water well pump, from* Groundwater Hydrology, *by H. Bouwer, 1978. Used with permission of McGraw-Hill Book Company.*

rangement when the static lift of a system is small relative to the total head loss due to friction and appurtenances at the design capacities. For this type of system, both the pump head and the system head vary proportionally with the square of the rotative speed. Consequently, a multiple-speed pump chosen to operate efficiently at one discharge will operate at acceptable efficiencies at other discharges as well.

EXAMPLE PROBLEM 5-6: Two pumps are operating in parallel as shown in Fig. 5-21. The head loss from 1 to 2 in the suction and discharge lines for each of the pumps can be determined using the equation

$$h_l = \frac{Q_i^2}{7,000,000} \tag{5-17}$$

where h_l = head loss in feet, and Q_i = individual pump discharge in gpm. System-head curves for the piping system beyond point 2 corresponding to new pipe and old pipe are given by

$$h_{new} = 20 + Q^2/196,000 \tag{5-18}$$

$$h_{old} = 20 + Q^2/140,000 \tag{5-19}$$

respectively, where h_{new} and h_{old} are in feet and Q is in gpm.

Determine the capacities and efficiencies of the pumps for each of the three operational modes for the "new" and "old" system curves.

Solution:

1. Determine modified pump curves.

    In analyzing parallel pumping systems it is recommended that *modified pump curves* be constructed by subtracting the head loss for the individual suction and discharge lines from the corresponding pump curves. The modified pump curves for pumps A and B are, therefore, determined by subtracting head loss calculated from Eq. (5-17) from the actual pump curves. These modified curves are shown as dashed lines in Fig. 5-21.

2. Develop composite modified pump curves and plot together with system curves and individual modified pump curves.

    The composite modified pump curve relating the head beyond point 2 for both pumps is obtained by adding the discharges from the modified pump curves for each pump at several heads. This curve, as well as the individual modified pump curves, is shown in Fig. 5-22. The system curves, given by Eqs. (5-18) and (5-19), are also presented in Fig. 5-22.

3. Determine operating conditions for both pumps with new pipe.

    The discharge for both pumps in operation when the system is new is 8,350 gpm at a head of 57.0 ft. The division of flow for this condition is determined by reading the discharge for pumps A and B from the modified pump curves in Fig. 5-21 for $h_p = 57$ ft. The values are $Q_A = 3,600$ gpm at 82.8% efficiency and $Q_B = 4,750$ gpm at 83.0% efficiency. The actual heads for pumps A and B, as read from the original pump curves in Fig. 5-21 are 58.2 and 60.1 ft, respectively.

4.   Determine operating conditions for other five modes.

The same general procedure is followed to determine the operational conditions for the system and for the individual pumps for the other five intersection points in Fig. 5-22. Table 5-3 summarizes the operational conditions corresponding to each of the six intersection points.

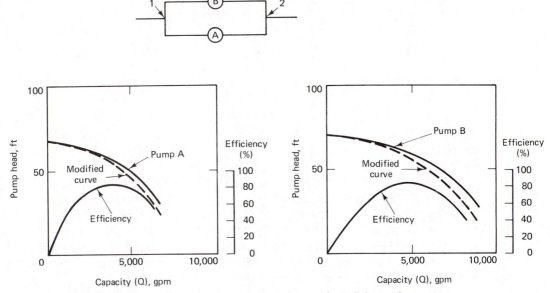

FIGURE 5-21   *Pump curves for individual pumps of parallel pumping system (Example Problem 5-6).*

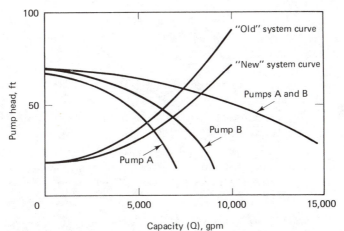

FIGURE 5-22   *Modified pump curves for three operational modes and system curves for "new" and "old" pipe roughness conditions (Example Problem 5-6).*

TABLE 5-3 *Operational conditions for intersection points in Fig. 5-22 of Example Problem 5-6.*

Operational mode	New pipe			Old pipe		
	A & B	A	B	A & B	A	B
$Q_A + Q_B$ (cfs)	8,350	5,730	6,720	7,380	5,400	6,220
h_p (ft)	57.0	37.2	43.4	59.4	40.8	48.0
Q_A (cfs)	3,600	5,730		3,000	5,400	
Q_B (cfs)	4,750		6,720	4,380		6,220
Eff_A (%)	82.8	68.5		80.0	73.2	
Eff_B (%)	83.0		70.2	82.4		77.2
h_A (ft)	58.2	40.8		60.7	45.0	
h_B (ft)	60.1		50.0	62.1		53.5

The analysis illustrated in Example Problem 5-6 should be undertaken when selecting pumps for a system requiring parallel pumping operation. In addition to ensuring that acceptable efficiencies are realized for all operational modes, the required NPSH must be met for all situations. If the static head varies significantly, operational conditions for system-head curves corresponding to minimum, maximum, and average static head conditions should be examined. For instance, if the static head in Example Problem 5-6 varied from 5 to 15 ft, the minimum, maximum, and average system-head curves for the new pipe and the corresponding intersection points are shown in Fig. 5-23. The operating conditions for these points can be found in the manner discussed in Example Problem 5-6.

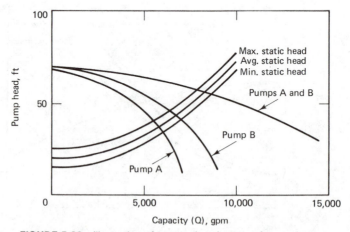

FIGURE 5-23 *Illustration of system-head curves for maximum, minimum, and average static heads for pumping arrangement of Example Problem 5-6.*

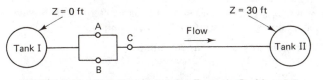

FIGURE 5-24 *Definition sketch for Example Problem 5-7.*

EXAMPLE PROBLEM 5-7: A pump with a 4-in.-diameter impeller has the following pump curve when operated at 800 rpm.

Q_0 (gpm)	0	400	800	1,200	1,600	2,000	2,400	2,800	3,000
h_{p_0} (ft)	62.5	62.7	62.5	61.2	57.7	52.4	43.6	30.8	19.0

This pump design is used for all three pumps in the multiple-pump system shown in Fig. 5-24. Pumps A and B have 5-in. impellers and pump C has a 6-in. impeller. Each pump is operated at 800 rpm. If the system curve, neglecting losses in the suction and discharge lines of the loop, is given by

$$h = 30 + \frac{Q^2}{370,000} \qquad (5\text{-}20)$$

where h = head in feet and Q = total discharge in gpm, determine the actual operating capacity and pumping head of the system.

Solution:

1. Using the affinity laws develop similarity equations relating pump head and discharge to rotative speed and impeller diameter.

    (a) Since ζ_{h_p} is constant for geometrically similar pumps, Eq. (5-6) can be used to develop the relationship

    $$\frac{h_{p_2}}{h_{p_1}} = \frac{(D_2^2 n_2^2)}{(D_1^2 n_1^2)} \qquad (5\text{-}21)$$

    Similary, Eq. (5-7) can be used to derive the equation

    $$\frac{Q_2}{Q_1} = \frac{(D_2^3 n_2)}{(D_1^3 n_1)} \qquad (5\text{-}22)$$

    (b) Similarity equations, based on Eqs. (5-21) and (5-22), expressing the pump head and capacity of pump A in terms of the pump for which the pump curve is given are

    $$h_{p_A} = \frac{(D_A^2 n_A^2)}{(D_0^2 n_0^2)} h_{p_0} = \frac{(5^2 \times 800^2)}{(4^2 \times 1000^2)} h_{p_0} = \mathbf{h_{p_0}}$$

    $$Q_A = \frac{(D_A^3 n_A)}{(D_0^3 n_0)} Q_0 = \frac{(5^3 \times 800)}{(4^3 \times 1000)} Q_0 = \mathbf{1.56 Q_0}$$

where h_{p_0}, Q_0, D_0, and n_0 are the parameters for the known pump. The equations are also applicable for h_{p_B} and Q_B.

(c) The similarity equations for pump C are

$$h_{p_C} = \frac{(6^2 \times 800^2)}{(4^2 \times 1000^2)} h_{p_0} = \mathbf{1.44} h_{p_0}$$

$$Q_c = \frac{(6^3 \times 800)}{(4^3 \times 1000)} Q_0 = \mathbf{2.70} Q_0$$

2. Determine the pump curve for pumps A, B, and C using the equations developed in Steps 1(b) and (c).

   (a) Applying the equations of Step 1(b) to the given pump curve yields the corresponding pump curve for pump A (and pump B) given in Columns 3 and 4 of Table 5-4.

   (b) Applying the equations of Step 1(c) to the given pump curve yields the corresponding pump curve for pump C given in Columns 5 and 6 of Table 5-4.

3. Determine the combined pump curve for pumps A and B operating in parallel.

   This is carried out by simply doubling discharges of pump A in Column 3 with the head being unchanged. The results are shown in Columns 7 and 8 of Table 5-4.

4. Determine the combined pump curve for pumps A, B, and C operating concurrently.

   Since the parallel pumps and pump C operate in series, this is accomplished by adding the ordinates of the combined pump curve for pumps A and B and the pump curve for pump C. The respective tabular pump curves given in Columns 7 and 8 and Columns 5 and 6 of Table 5-4 have

TABLE 5-4 Pump curves for Example Problem 5-7.

Q_0 (gpm) (1)	h_{p_0} (ft) (2)	Q_A (gpm) (3)	h_{p_A} (ft) (4)	Q_C (gpm) (5)	h_{p_C} (ft) (6)	Q_{A+B} (gpm) (7)	$h_{p_{A+B}}$ (ft) (8)
0	62.5	0	62.5	0	90.0	0	62.5
400	62.7	625	62.7	1,080	90.3	1,250	62.7
800	62.5	1,250	62.5	2,160	90.0	2,500	62.5
1,200	61.2	1,875	61.2	3,240	88.1	3,750	61.2
1,600	57.7	2,500	57.7	4,320	83.1	5,000	57.7
2,000	52.4	3,125	52.4	5,400	74.5	6,250	52.4
2,400	43.6	3,750	43.6	6,480	62.8	7,500	43.6
3,200	19.0	5,000	19.0	8,640	27.4	10,000	19.0

different discharges, consequently, this procedure is best carried out graphically as illustrated in Fig. 5-25.

5. Determine the actually operating conditions.

> The system curve [Eq. (5-29)] is plotted on Fig. 5-25. The intersection of this curve with the combined pump curve for pumps A, B, and C operating together is the operating point. The capacity and head corresponding to this point are

$$Q = \textbf{5,800 gpm}$$

and

$$h = \textbf{123 ft}$$

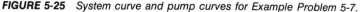

FIGURE 5-25   *System curve and pump curves for Example Problem 5-7.*

REFERENCES

BOUWER, H., *Groundwater Hydrology,* 1st Edition, McGraw–Hill Book Company, New York, New York, (1978).

DAUGHERTY, R. L., and FRANZINI, J. B., *Fluid Mechanics with Engineering Applications,* 7th Edition, McGraw–Hill Book Company, New York, New York (1977).

HWANG, N. H. C., *Fundamentals of Hydraulic Engineering Systems,* 1st Edition, Prentice-Hall, Englewood Cliffs, New Jersey (1981).

KARASSIK, I. J., and CARTER, R., *Centrifugal Pumps,* 1st Edition, F. W. Dodge Corporation, New York, New York (1960).

KARASSIK, I. J., KRUTZSCH, W. C., FRASER, W. H., and MESSINA, J. P., *Pump Handbook,* 1st Edition, McGraw–Hill Book Company, New York (1976).

STREETER, V. L., and WYLIE F. B., Fluid Mechanics, 7th Edition, McGraw–Hill Book Company, New York, New York (1979).

TCHOBANOGLOUS, G., *Wastewater Engineering: Collection and Pumping of Wastewater,* 1st Edition, McGraw–Hill Book Company, New York, New York (1981).

6

DESIGN EXAMPLE FOR A WASTEWATER TREATMENT PLANT

After the process design of a treatment plant has been completed, i.e., the number and size of the individual process units have been determined, the hydraulic design must be undertaken. This means that the spatial arrangement of the individual units must be established and each unit connected by pipes and/or channels which are sized and located so that hydraulic compatibility is achieved. Water surface elevations are generally calculated for the maximum and average flows, which are expected to occur at the end of the design period, as well as for the minimum flow initially expected. Furthermore, these calculations should cover the range of operational conditions expected to occur, e.g., when all units are operational or when specific units are out of service because of maintenance requirements or because they are not needed during the early years of plant operation. The results of the water surface elevation calculations are usually summarized in the form of a hydraulic profile for each flow scenario (see Fig. 6-1). As noted by Metcalf and Eddy (1979), these profiles are used to: (a) ensure that gravity flow is possible between each process unit, (b) provide the information necessary to calculate the head requirement when pumping is needed, and (c) ensure that the plant does not back up or flood under peak flow conditions.

The hydraulic design of a treatment plant is a trial-and-error process where the size of individual hydraulic appurtenances are initially assumed and a particular flow scenario (flow rate and number of operational units) is established. A flow analysis is then conducted where a number of factors are evaluated. These include: (a) the water surface elevation, (b) the magnitude of the energy loss in each component, (c) the flow distribution in dividing-flow manifolds and distribution channels, and (d) the velocity of flow in each component.

Available head is a precious commodity at most treatment plants and should be given up grudgingly. If the water surface elevation computations indicate that the energy loss in a particular appurtenance is excessive, its size may have to be increased so that the energy loss will be reduced. Fairly even flow distribution to parallel units is also important for high process efficiency. If the hydraulic analysis indicates a

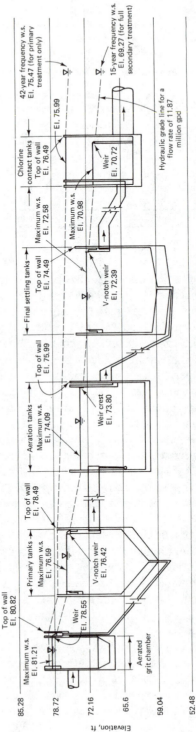

FIGURE 6-1 Hydraulic profile for Westfield, Massachusetts Wastewater Treatment Plant, from Wastewater Engineering: Collection, Treatment, Disposal, by Metcalf and Eddy, Inc., 1979. Used with permission of McGraw-Hill Book Company.

172

significant variation in the flow distribution, a size adjustment may be necessary in the main pipe or channel or in the control devices such as the orifices or weirs used to distribute the flow. Flow velocity is also an important consideration in hydraulic design. The ideal situation is to size the interconnecting conduits to prevent solids' deposition under all flow conditions. This usually means velocities greater than 2.0 ft/s at design average flow and velocities greater than 1.0 ft/s at the initial minimum flow. In many cases such values may be impossible to achieve, but the flushing action provided by the high-flow period experienced during the diurnal flow variation may be depended upon to keep the conduits cleaned. In other cases mixing by aeration is often used to prevent solids' deposition. In this regard, mixing by aeration is commonly employed in distribution channels where downstream velocities may be quite low even at maximum flow.

The sole purpose of this chapter is to illustrate many of the calculations required to establish the water surface elevation through a wastewater treatment plant. The plant layout employed for these calculations should, in no way, be interpreted as ideal. A flow schematic is used, which requires several different types of calculations rather than being hydraulically efficient or even practical. The computational steps are presented in detail and many of them are repetitious. Such an approach, although tedious, is less confusing and much easier for someone unfamilar with the procedure to follow.

The computations presented in subsequent sections are based on the flow schematic presented in Fig. 6-2. The sizes of the individual units shown in this figure are given in Table 6-1. The hydraulic analysis will be based on one particular situation where a flow of 24.9 cfs with all units in operation is assumed. Again, it is emphasized

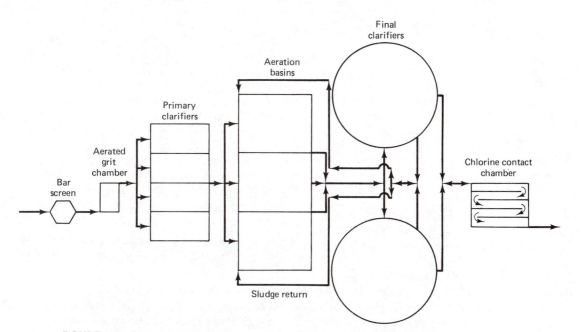

FIGURE 6-2 *Flow schematic for treatment plant to be used in hydraulic analysis.*

TABLE 6-1 *Design summary for treatment plant to be used in hydraulic analysis.*

Item	Value
Influent sewer invert elevation, ft	300.00
Influent sewer size, in.	36
Wastewater flow rate, cfs	24.9
Aerated grit chamber	
Number of units	1
Length of units, ft	26.4
Width of unit, ft	18.0
Depth of unit, ft*	9.0
Primary clarifier, rectangular	
Number of units	4
Length of unit, ft	100.0
Width of unit, ft	25.0
Depth of unit, ft	9.0
Aeration tanks, rectangular	
Number of units	3
Length of unit, ft	156.0
Width of unit, ft	52.0
Depth of unit, ft*	10.0
Sludge recirculation ratio	0.55
Final clarifier, circular	
Number of units	2
Diameter of unit, ft	105.0
Depth of unit, ft	14.0
Bottom slope, %	1.0
Chlorine contact chamber	
Number of units	1
Length of unit, ft	120.0
Width of unit, ft	43.0
Depth of unit, ft	8.4
Number of chambers	5
Width of chamber, ft	8.6

* If a diffused aeration system is used, the depth of the aerated grit chamber and aeration tanks should be the same to simplify air distribution.

that the material to be presented is intended only to illustrate typical calculations that the engineer has to make when evaluating the hydraulic characteristics of a plant. Furthermore, the computations should not be considered to constitute a complete hydraulic design. No attempt is made to adjust conduit sizes when velocities fall below the recommended minimum or when energy losses are excessive and only one flow scenario is studied.

There are many ways to begin the computations for any hydraulic analysis. The particular method selected depends on the preference of the design engineer as well as on local conditions. For example, the high water level in the receiving stream may be a control point. Many engineers will begin at this point and work back through the plant. On the other hand, certain engineers may prefer to select a control point in the center of the plant and work in both directions. The starting point selected for the hydraulic calculations presented in this chapter will be the control weir for the aerated grit chamber. Hydraulic computations will proceed in both directions from this point.

It is common practice in treatment plant design to place flow channels on zero slope. Thus, the water surface elevation between two points in a straight channel, where energy losses can be neglected, will be constant.

6.1 HYDRAULIC ANALYSIS FOR GRIT CHAMBER AND UPSTREAM COMPONENTS

The effluent weir for the aerated grit chamber is the first control point in the treatment plant (see Fig. 6-3). Before water surface profiles can be evaluated, it is necessary to establish the locations of the plant *control points*. These are points where there is a definite relationship between discharge and depth. Hence, control points occur at weirs and at any point where critical depth occurs, e.g., at free overfalls. Water surface profile computations proceed upstream from a control point, if flow is subcritical, and downstream from a control point, if flow is supercritical. Flow in treatment plants is normally subcritical.

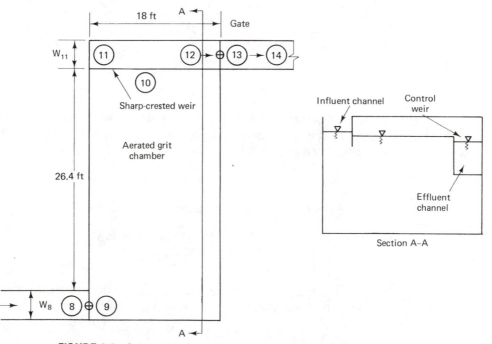

FIGURE 6-3 *Schematic of aerated grit chamber and ancillary hydraulic components.*

Calculations will proceed upstream from the control weir of the aerated grit chamber. These calculations consider minor energy losses but neglect friction losses. In all computations uniform flow is assumed. Although the actual flow conditions are generally gradually varied, the accuracy gained by following the numerical integration techniques of gradually varied flow is generally not worth the effort.

Computations

1. Assume an elevation for the channel bottom between the plant entrance and the entrance to the grit chamber. A reasonable first estimate for the channel bottom elevation can be made by assuming that it is at the same elevation as the invert of the influent sewer at the plant entrance (or the feed channel bottom elevation at the plant entrance if the wastewater enters the plant via a forced main). Thus, the channel bottom elevation (CBE) at point 8 shown on Fig. 6-3 will be assumed to be 300.00 ft.

$$CBE(8) = 300.00 \text{ ft}$$

2. Establish the water surface elevation at points 9 and 10 shown on Fig. 6-3.

   (a) Calculate the head on the control weir from Eq. (1-35).

$$H = \left[\frac{1}{(C_w)^{2/3}(L)^{2/3}} \right] Q^{2/3}$$

$$= \left[\frac{1}{(3.33)^{2/3}(18)^{2/3}} \right] (24.9)^{2/3}$$

$$= 0.57 \text{ ft}$$

   (b) Assume an elevation, EW1, for the grit chamber weir crest. This value should be slightly larger than the channel bottom elevation at the entrance to the grit chamber.

$$EW1 = 300.71 \text{ ft}$$

   (c) Establish the water surface elevation (WSE) at point 10.

$$Z_{10} = WSE(10) = 300.71 + 0.57$$

$$= 301.28 \text{ ft}$$

   (d) Neglect the energy loss due to friction through the grit chamber and establish the water surface elevation at point 9 shown on Fig. 6-3.

$$Z_9 = WSE(9) = WSE(10)$$

$$= 301.28 \text{ ft}$$

3. Assume a width for the approach channel to the grit chamber. This is shown as W_8 in Fig. 6-3.

$$W_8 = \mathbf{4.0 \ ft}$$

The assumed width should be compatible with the size of the equipment available for the bar screen channel.

4. For the specified width, W_8, calculate the critical depth that would produce a *choke* at the design flow rate at point 8 shown on Fig. 6-3. In this instance *choke* refers to a constriction that is severe enough to influence the upstream flow.

   Compute the critical depth for flow in a rectangular channel.

$$y_c = \left[\frac{(q)^2}{g} \right]^{1/3} = \left[\frac{Q^2}{gw^2} \right]^{1/3}$$

$$= \left[\frac{(24.9)^2}{(32.2)(4)^2} \right]^{1/3}$$

$$= \mathbf{1.064 \ ft}$$

5. Apply the energy equation between points 8 and 9 and establish the depth of flow at point 8.

$$Z_8 + \frac{V_8^2}{2g} = Z_9 + \frac{V_9^2}{2g} + K_{\text{gate}} \frac{V_8^2}{2g} + K_{\text{ent}} \frac{V_8^2}{2g}$$

where Z_8 and Z_9 represent the water surface elevations at points 8 and 9, respectively. By neglecting the velocity head in the grit chamber and letting the entrance and the gate recess loss coefficients equal 1.0 and 0.2, respectively, the following relationship is obtained;

$$Z_8 + \frac{V_8^2}{2g} = Z_9 + 1.2 \left[\frac{V_8^2}{2g} \right]$$

or

$$Z_8 + \left[\frac{Q^2}{2gw_8^2 y_8^2} \right] = Z_9 + 1.2 \left[\frac{Q^2}{2gw_8^2 y_8^2} \right]$$

Since

$$Z_8 = \text{CBE}(8) + y_8$$

the above expression can be rearranged to give

$$y_8 - 0.2\left[\frac{Q^2}{2gw_8^2 y_8^2}\right] = 301.28 - 300.00$$

$$y_8 - \frac{0.12}{y_8^2} = \mathbf{1.28}$$

This is a cubic equation with two real roots (provided the channel is not so narrow that a choke condition exists). The higher of the two roots corresponds to the subcritical flow condition. The root is found by iteration to be $y_8 = \mathbf{1.35\ ft.}$

6. Compare the depth computed in Step 5 to the critical depth computed in Step 4. The depth from Step 5 should be larger than the depth from Step 4 to avoid a choke condition. The depth of flow computed in Step 5 can be changed by adjusting the grit chamber control weir elevation. When evaluating the acceptability of the flow depth at point 8, not only should the choke condition be considered, but it should also be remembered that there are certain restrictions on the velocity through the bar screen, which is located upstream of point 8. The "Ten-State Standards" limits the velocity through the bar openings to 2.5 fps. However, velocities of 2 to 4 fps through the openings have been used. The smaller the depth of flow at point 8 the larger will be the upstream velocities.

7. Compute the water surface elevation at point 8.

$$\text{WSE}(8) = \text{CBE}(8) + y_8$$

$$= 300.00 + 1.35$$

$$= \mathbf{301.35\ ft}$$

8. Calculate the water surface elevation and depth of flow at point 7 shown on Fig. 6-4. Neglect the energy loss due to friction between points 7 and 8.

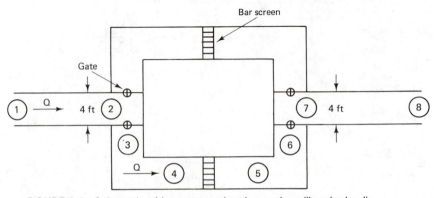

FIGURE 6-4 Schematic of bar screen chamber and ancillary hydraulic components.

$$Z_7 = \text{WSE}(7) = \text{WSE}(8)$$

$$= \mathbf{301.35 \ ft}$$

$$y_7 = \text{WSE}(8) - \text{CBE}(8)$$

$$= 301.35 - 300.00$$

$$= \mathbf{1.35 \ ft}$$

9. Apply the energy equation between points 6 and 7 and establish the depth of flow at point 6.

$$y_6 + \frac{V_6^2}{2g} = y_7 + \frac{V_7^2}{2g} + K_{90}\frac{V_6^2}{2g} + K_{\text{gate}}\frac{V_6^2}{2g}$$

Since

$$V_6 = \frac{Q_6}{A_6} = \frac{Q_6}{y_6 w_6}$$

The energy equation may be rearranged into the form

$$y_6 + (1 - K_{90} - K_{\text{gate}})\frac{(24.9)^2}{2gw_6^2 y_6^2} = 1.35 + \frac{(24.9)^2}{(64.4)(4)^2(1.35)^2}$$

Assuming a value of 0.3 for K_{90} and a value of 0.2 for K_{gate}, it is possible to write

$$y_6 + (1 - 0.3 - 0.2)\frac{(24.9)^2}{(64.4)(4)^2 y_6^2} = 1.35 + \frac{(24.9)^2}{(64.4)(4)^2(1.35)^2}$$

$$y_6 + \frac{0.30}{y_6^2} = 1.68$$

The solution of this equation is found to be $y_6 = \mathbf{1.56 \ ft}$.

10. Apply the energy equation between point 5 and point 6 and establish the depth of flow at point 5.

$$y_5 + \frac{V_5^2}{2g} = y_6 + \frac{V_6^2}{2g} + K_{90}\frac{V_5^2}{2g}$$

or

$$y_5 + (1 - K_{90})\frac{Q_5^2}{2gw_5^2 y_5^2} = y_6 + \frac{Q_6^2}{2gw_6^2 y_6^2}$$

Assuming a value of 0.3 for K_{90} gives:

$$y_5 + (1 - 0.3)\frac{(24.9)^2}{(64.4)(4)^2 y_5^2} = 1.56 + \frac{(24.9)^2}{(64.4)(4)^2(1.56)^2}$$

$$y_5 + \frac{0.42}{y_5^2} = 1.81$$

The solution of this equation is found to be $y_5 = $ **1.66 ft.**

11. Calculate the water surface elevation at point 5 shown on Fig. 6-4.

$$\text{WSE}(5) = \text{CBE}(5) + y_5$$

$$= 300.00 + 1.66$$

$$= \textbf{301.66 ft}$$

Note: Throughout this chapter all flow channels will be placed on zero slope.

12. Assuming that the bar screen will be mechanically cleaned when the drop in the water surface elevation across the screen reaches 6 in., establish the water surface elevation and depth of flow at point 4 shown on Fig. 6-4.

$$\text{WSE}(4) = \text{WSE}(5) + 0.5$$

$$= 301.66 + 0.5$$

$$= \textbf{302.16 ft}$$

$$y_4 = \text{WSE}(4) - \text{CBE}(4)$$

$$= 302.16 - 300.00$$

$$= \textbf{2.16 ft}$$

13. Check the average velocity of flow in the bar screen chamber to ensure that it is acceptable.

$$V_4 = \frac{Q_4}{A_4}$$

$$= \frac{24.9}{(4)(2.16)}$$

$$= \textbf{2.88 fps}$$

The velocity in the screen chamber should be greater than 1.3 fps at minimum flows to avoid grit deposition when grit chambers follow bar screens. Assuming these calculations are based on peak flow, the velocity of 2.88 fps is adequate. The velocity could be changed, however, by assuming a different channel width (Step 3), or by selecting a different grit chamber control weir elevation (Step 2).

14. Apply the energy equation between points 3 and 4 and establish the depth of flow at point 3.

$$y_3 + \frac{V_3^2}{2g} = y_4 + \frac{V_4^2}{2g} + K_{90}\frac{V_3^2}{2g}$$

or

$$y_3 + (1 - K_{90})\frac{Q_3^2}{2gw_3^2y_3^2} = y_4 + \frac{Q_4^2}{2gw_4^2y_4^2}$$

assuming a value of 0.3 for K_{90} gives

$$y_3 + (1 - 0.3)\frac{(24.9)^2}{(64.4)(4)^2y_3^2} = 2.16 + \frac{(24.9)^2}{(64.4)(4)^2(2.16)^2}$$

$$y_3 + \frac{0.42}{y_3^2} = 2.29$$

The solution of this equation is found to be $y_3 = $ **2.20 ft.**

15. Apply the energy equation between points 2 and 3 and establish the depth of flow at point 2.

$$y_2 + \frac{V_2^2}{2g} = y_3 + \frac{V_3^2}{2g} + K_{90}\frac{V_2^2}{2g} + K_{gate}\frac{V_2^2}{2g}$$

or

$$y_2 + (1 - K_{90} - K_{gate})\frac{Q_2^2}{2gw_2^2y_2^2} = y_3 + \frac{Q_3^2}{2gw_3^2y_3^2}$$

Assuming a value of 0.3 for K_{90} and a value of 0.2 for K_{gate} gives

$$y_2 + (1 - 0.3 - 0.2)\frac{(24.9)^2}{(64.4)(4)^2y_2^2} = 2.20 + \frac{(24.9)^2}{(64.4)(4)^2(2.20)^2}$$

$$y_2 + \frac{0.30}{y_2^2} = 2.33$$

The solution of this equation is found to be $y_2 = $ **2.27 ft.**

16. Compute the water surface elevation and the depth of flow at point 1 shown on Fig. 6-4.

$$\text{WSE}(1) = \text{CBE}(2) + y_2$$
$$= 300.00 + 2.27$$
$$= 302.27 \text{ ft}$$
$$y_1 = y_2$$
$$= \mathbf{2.27 \text{ ft}}$$

17. Check the average velocity at point 1.

$$V_1 = \frac{Q_1}{A_1} = \frac{Q_1}{w_1 y_1}$$
$$= \frac{24.9}{(4)(2.27)}$$
$$= \mathbf{2.74 \text{ fps}}$$

Acceptable practice is to provide a flow velocity of 2.0 fps at average flow.

18. Check the influent sewer to ensure that it is not surcharged.

$$\begin{bmatrix} \text{Elevation at crown} \\ \text{of influent sewer} \end{bmatrix} = \begin{bmatrix} \text{invert} \\ \text{elevation} \end{bmatrix} + \begin{bmatrix} \text{sewer} \\ \text{diameter} \end{bmatrix}$$
$$= 300.00 + 3.00$$
$$= \mathbf{303.00 \text{ ft}}$$

Since

$$303.00 > 302.27$$

the influent sewer will not be surcharged. Had the water surface elevation at point 1 been greater than or nearly equal to the elevation at the crown of the influent sewer, the channel bottom elevation and the grit chamber control weir elevation would have had to be lowered.

6.2 HYDRAULIC ANALYSIS FOR PRIMARY CLARIFIER SYSTEM

The next control point is located at the effluent weir of the primary clarifiers. The elevation of this weir will control the water surface elevation between point 16 on Fig. 6-5 and point 11 on Fig. 6-3, i.e., between the first and second control points. To make

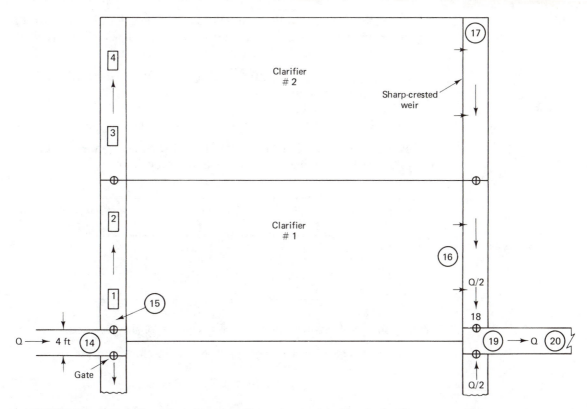

FIGURE 6-5 *Schematic of the primary clarifier system and ancillary hydraulic components.*

the water surface elevation calculations between these two control points, it is first necessary to establish the control weir elevation for the clarifier system and the elevation of the distribution channel bottom. Once these parameters are set, the flow distribution to the clarifiers and the depth of flow in the distribution channel can be determined by following the computational technique for distribution channels outlined in Chap. 3. When the depth of flow at point 15 is found, the depth of flow at point 12 can be established by making calculations similar to those carried out for the bar screen channel. After the depth of flow at point 12 is computed, the depth of flow at point 11 is calculated from Eq. (3-50). The water surface elevation at point 11 is then determined and compared to the control weir elevation for the aerated grit chamber. The water surface at point 11 should be at least 2 in. below the control weir elevation to permit the weir to behave as a freely discharging weir.

Computations

1. As a first approximation assume that the control weir elevation for the clarifiers is 297.80 ft and that the distribution channel bottom is 6 in. below the control weir, i.e., the elevation of the distribution channel bottom is 297.30 ft.

2. Establish the number of orifices to be used, the spacing of the orifices, the size of the orifices, and the width of the distribution channel. For the problem under consideration the following values will be assumed:

> Orifice: 2 ft × 0.5 ft, rectangular, sharp edged
> Channel width: 4 ft
> Number of orifices: 4 per channel leg
> Orifice spacing: 12.5 ft, center to center

3. Following the computational procedure for distribution channels outlined in Chap. 3, the flow parameter values of interest are

Orifice no.	Flow (cfs)	Depth of flow (ft)	Velocity of flow (ft/s)
4	3.136	2.085	0.377
3	3.132	2.080	0.754
2	3.125	2.071	1.135
1	3.114	2.058	1.520

Note: In actual practice more orifices would probably be used.

4. Establish the water surface elevation at point 15 shown on Fig. 6-5.

$$\text{WSE}(15) = \text{CBE}(15) + y_{15}$$

$$= 297.30 + 2.06$$

$$= \textbf{299.36 ft}$$

Note: No allowance will be made for the energy loss through gate recesses in the middle of lateral spillway or distribution channels. It is normally small and not worth the effort required to incorporate it in the analysis.

5. Apply the energy equation between points 14 and 15 and establish the depth of flow at point 14.

$$y_{14} + \frac{V_{14}^2}{2g} = y_{15} + \frac{V_{15}^2}{2g} + K_{90}\frac{V_{14}^2}{2g} + K_{\text{gate}}\frac{V_{14}^2}{2g}$$

or

$$y_{14} + (1 - K_{90} - K_{\text{gate}})\frac{Q_{14}^2}{2gw_{14}^2 y_{14}^2} = y_{15} + \frac{Q_{15}^2}{2gw_{15}^2 y_{15}^2}$$

Assuming a value of 0.3 for K_{90} and a value of 0.2 for K_{gate}

$$y_{14} + (1 - 0.3 - 0.2)\frac{(24.9)^2}{(64.4)(4)^2 y_{14}^2} = 2.06 + \frac{(24.9/2)^2}{(64.4)(4)^2(2.06)^2}$$

$$y_{14} + \frac{0.30}{y_{14}^2} = 2.10$$

The solution of this equation is found to be $y_{14} = $ **2.03 ft.**

6. Establish the water surface elevation at point 14 shown on Figs. 6-5 and 6-3.

$$WSE(14) = CBE(14) + y_{14}$$
$$= 297.30 \text{ ft} + 2.03 \text{ ft}$$
$$= \mathbf{299.33 \text{ ft}}$$

7. Assuming no energy loss due to friction, establish the water surface elevation and the depth of flow at point 13 shown on Fig. 6-3.

$$WSE(13) = WSE(14)$$
$$= \mathbf{299.33 \text{ ft}}$$

$$y_{13} = y_{14}$$
$$= \mathbf{2.03 \text{ ft}}$$

8. Apply the energy equation between points 12 and 13 shown on Fig. 6-3 and establish the depth of flow at point 12.

$$y_{12} + \frac{V_{12}^2}{2g} = y_{13} + \frac{V_{13}^2}{2g} + K_{gate}\frac{V_{12}^2}{2g}$$

or

$$y_{12} + (1 - K_{gate})\frac{Q_{12}^2}{2gw_{12}^2 y_{12}^2} = y_{13} + \frac{Q_{13}^2}{2gw_{13}^2 y_{13}^2}$$

Assuming a value of 0.2 for K_{gate}

$$y_{12} + (1 - 0.2)\frac{(24.9)^2}{(64.4)(4)^2 y_{12}^2} = 2.03 + \frac{(24.9)^2}{(64.4)(4)^2(2.03)^2}$$

$$y_{12} + \frac{0.48}{y_{12}^2} = 2.18$$

The solution of this equation is found to be $y_{12} = $ **2.07 ft.**

9. Establish the water surface elevation at point 12.

$$WSE(12) = CBE(12) + y_{12}$$
$$= 297.30 + 2.07$$
$$= \mathbf{299.37 \text{ ft}}$$

10. Determine the depth of flow at point 11 on Fig. 6-3. This can be accomplished by applying the principles developed in the *Lateral Spillway Channels* segment of Sec. 3.2 of this text. In this case the necessary computations can be made by applying Eq. (3-50).

$$y_u = \left[\frac{2(y_c)^3 + (y_l)^3}{y_l} \right]^{1/2}$$

(a) Calculate the critical depth from Eq. (3-18).

$$y_c = \left[\frac{Q^2}{gw^2} \right]^{1/3}$$

$$= \left[\frac{(24.9)^2}{(32.2)(4)^2} \right]^{1/3}$$

$$= 1.06 \text{ ft}$$

(b) Compute the depth of flow at point 11 from Eq. (3-50).

$$y_{11} = \left[\frac{2(1.06)^3 + (2.07)^3}{2.07} \right]^{1/2}$$

$$= 2.33 \text{ ft}$$

11. Establish the water surface elevation at point 11.

$$\text{WSE}(11) = \text{CBE}(11) + y_{11}$$

$$= 297.30 + 2.33$$

$$= 299.63 \text{ ft}$$

12. Compare the water surface elevation at point 11 to the elevation of the control weir for the aerated grit chamber. This comparison is necessary to ensure that there is at least 2 in. of free-fall from the control weir to the lateral spillway channel.

Control weir elevation: 300.71 ft

WSE(11): 299.63 ft

Difference: 1.08 ft

This difference is excessive. It should be reduced by raising the elevation of the primary clarifier system control weir by an amount equal to $1.08 - \frac{2}{12} =$ **0.91 ft.**

13. Establish the water surface profile between the primary clarifiers and aerated grit chamber based on a 298.71-ft elevation for the primary clarifier system

control weir and an elevation of $298.71 - 0.5 = 298.21$ ft for the primary
clarifier system distribution channel bottom.

$$WSE(15) = 299.36 \text{ ft} + 0.91 \text{ ft} = 300.27 \text{ ft}$$
$$WSE(14) = 199.33 \text{ ft} + 0.91 \text{ ft} = 300.24 \text{ ft}$$
$$WSE(13) = 299.33 \text{ ft} + 0.91 \text{ ft} = 300.24 \text{ ft}$$
$$WSE(12) = 299.37 \text{ ft} + 0.91 \text{ ft} = 300.28 \text{ ft}$$
$$WSE(11) = 299.63 \text{ ft} + 0.91 \text{ ft} = 300.54 \text{ ft}$$

6.3 HYDRAULIC ANALYSIS FOR AERATION BASIN SYSTEM

The third control point for the system illustrated in Fig.6-2 is located at the effluent
weir of the aeration basin system. The elevation of this weir will control the water
surface elevation between point 24 on Fig.6-6 and point 17 on Fig. 6-5, i.e., between
the second and third control points. To make the water surface elevation calculations
between these two control points, it is first necessary to establish the control weir
elevation for the aeration basins and the elevation of the distribution channel bottom.

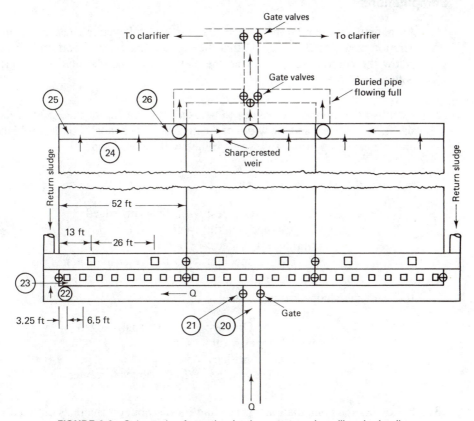

FIGURE 6-6 *Schematic of aeration basin system and ancillary hydraulic
components.*

Once these parameters are set, the flow distribution to the basins and the depth of flow in the distribution channel can be determined by following the computational procedure for distribution channels outlined in Chap. 3. When the depth of flow at point 23 is found, the depth of flow at point 18 can be established by making calculations similar to those carried out for the bar screen channel. After the depth of flow at point 18 is computed, the depth of flow at point 17 is calculated from Eq. (3-50). The water surface elevation at point 17 is then determined and compared to the control weir elevation for the primary clarifier system. The water surface elevation at point 17 should be at least 2 in. below the control weir elevation.

The channel arrangement from the primary clarifiers to the aeration basins provides a good deal of flexibility in flow control. This arrangement allows the aeration basins to be fed from the left side, the right side, or both sides. This means that any of the aeration basins may be taken off-line for maintenance without dramatically affecting the flow distribution. The same flexibility is built into the return sludge system (see Fig. 6-6).

Computations

1. As a first approximation assume that the control weir elevation for the aeration basins is 295.3 ft and that the distribution channel bottom is 6 in. below the control weir, i.e., that the elevation of the distribution channel bottom is 294.8 ft.

2. Establish the number of orifices to be used, the spacing of the orifices, the size of the orifices, and the width of the distribution channel. For the problem under consideration the following values will be assumed:

    (a) For the wastewater channel

Orifices:	0.5 ft^2 sharp edged
Channel width:	4 ft
Number of orifices:	24
Orifice spacing:	6.5 ft, center to center

    (b) For the sludge return channel

Orifices:	0.75 ft^2 sharp edged
Channel width:	4 ft
Number of orifices:	6
Orifice spacing:	26-ft, center to center

3. Following the computational procedure for distribution channels outlined in Chap. 3, the flow parameter values of interest are:

Orifice no.	Flow (cfs)	Depth of flow (ft)	Velocity (ft/s)
24	1.053	1.460	0.182
23	1.053	1.460	0.361
22	1.053	1.460	0.541
21	1.053	1.460	0.721
20	1.053	1.460	0.902
19	1.053	1.460	1.082
18	1.053	1.460	1.262
17	1.053	1.460	1.443
16	1.055	1.460	1.623
15	1.054	1.460	1.804
14	1.054	1.460	1.984
13	1.053	1.460	2.165
12	1.052	1.460	2.345
11	1.051	1.460	2.525
10	1.049	1.460	2.705
9	1.045	1.460	2.884
8	1.043	1.460	3.062
7	1.040	1.460	3.241
6	1.037	1.460	3.418
5	1.032	1.460	3.594
4	1.028	1.460	3.771
3	1.022	1.460	3.946
2	1.016	1.460	4.121
1	1.009	1.460	4.294

For the sludge return channel

Orifice no.	Flow (cfs)	Depth of flow (ft)	Velocity (ft/s)
6	2.335	1.358	0.439
5	2.318	1.351	0.862
4	2.302	1.339	1.300
3	2.273	1.322	1.747
2	2.236	1.299	2.208
1	2.178	1.269	2.690

4. Establish the water surface elevation at point 23 shown on Fig. 6-6.

$$WSE(23) = CBE(23) + y_{23}$$
$$= 294.80 + 1.46$$
$$= \mathbf{296.26 \ ft}$$

5. Apply the energy equation between points 22 and 23 shown on Fig. 6-6 and establish the depth of flow at point 22.

$$y_{22} + \frac{V_{22}^2}{2g} = y_{23} + \frac{V_{23}^2}{2g} + K_{gate}\frac{V_{22}^2}{2g} + K_{180}\frac{V_{22}^2}{2g}$$

or

$$y_{22} + (1 - K_{gate} - K_{180})\frac{Q_{22}^2}{(64.4)w_{22}^2 y_{22}^2} = y_{23} + \frac{Q_{23}^2}{(64.4)w_{23}^2 y_{23}^2}$$

Assume a value of 0.2 for K_{gate} and approximate the energy loss due to the 180° bend by assuming that it is nearly the same as the loss through two 90° bends where K is normally taken as 0.3. Then

$$y_{22} + (1 - 0.2 - 0.6)\frac{(24.9)^2}{(64.4)(4)^2 y_{22}^2} = 1.46 + \frac{(24.9)^2}{(64.4)(4)^2(1.46)^2}$$

$$y_{22} + \frac{0.12}{y_{22}^2} = 1.74$$

The solution of this equation is found to be $y_{22} = $ **1.70 ft.**

6. Establish the water surface elevation at point 22.

$$WSE(22) = CBE(22) + y_{22}$$
$$= 294.80 + 1.70$$
$$= \textbf{296.50 ft}$$

7. Assuming no energy loss due to friction, establish the water surface elevation and depth of flow at point 21 shown on Fig. 6-6.

$$WSE(21) = WSE(22)$$
$$= 296.50 \text{ ft}$$
$$y_{21} = y_{22}$$
$$= \textbf{1.70 ft}$$

8. Apply the energy equation between points 20 and 21 shown on Fig. 6-6 and establish the depth of flow at point 20.

$$y_{20} + \frac{V_{20}^2}{2g} = y_{21} + \frac{V_{21}^2}{2g} + K_{gate}\frac{V_{20}^2}{2g} + K_{90}\frac{V_{20}^2}{2g}$$

or

$$y_{20} + (1 - K_{\text{gate}} - K_{90}) \frac{Q_{20}^2}{(64.4)w_{20}^2 y_{20}^2} = y_{21} + \frac{Q_{21}^2}{(64.4)w_{21}^2 y_{21}^2}$$

Assuming a value of 0.2 for K_{gate} and 0.3 for K_{90}

$$y_{20} + (1 - 0.2 - 0.3) \frac{(24.9)^2}{(64.4)(4)^2 y_{20}^2} = 1.70 + \frac{(24.9)^2}{(64.4)(4)^2(1.70)^2}$$

$$y_{20} + \frac{0.30}{y_{20}^2} = 1.91$$

The solution to this equation is found to be $y_{20} = $ **1.82 ft.**

9. Establish the water surface elevation at point 20.

$$\text{WSE}(20) = \text{CBE}(20) + y_{20}$$

$$= 294.80 + 1.82$$

$$= \textbf{296.62 ft}$$

10. Assuming no energy loss due to friction, establish the water surface elevation and the depth of flow at point 19 shown on Fig. 6-5.

$$\text{WSE}(19) = \text{WSE}(20)$$

$$= 296.62 \text{ ft}$$

$$y_{19} = y_{20}$$

$$= \textbf{1.82 ft}$$

11. Apply the energy equation between points 18 and 19 shown on Fig. 6-5 and establish the depth of flow at point 18.

$$y_{18} + \frac{V_{18}^2}{2g} = y_{19} + \frac{V_{19}^2}{2g} + K_{90} \frac{V_{18}^2}{2g} + K_{\text{gate}} \frac{V_{18}^2}{2g}$$

or

$$y_{18} + (1 - K_{90} - K_{\text{gate}}) \frac{Q_{18}^2}{(64.4)w_{18}^2 y_{18}^2} = y_{19} + \frac{Q_{19}^2}{(64.4)w_{19}^2 y_{19}^2}$$

Assuming a value of 0.2 for K_{gate} and 0.3 for K_{90}

$$y_{18} + (1 - 0.3 - 0.2) \frac{(24.9/2)^2}{(64.4)(4)^2 y_{18}^2} = 1.82 + \frac{(24.9)^2}{(64.4)(4)^2(1.82)^2}$$

$$y_{18} + \frac{0.075}{y_{18}^2} = 2.00$$

The solution to this equation is found to be y_{18} = **1.98 ft.**

12. Establish the water surface elevation at point 18.

$$\text{WSE}(18) = \text{CBE}(18) + y_{18}$$
$$= 294.80 + 1.98$$
$$= \textbf{296.78 ft}$$

13. Determine the depth of flow at point 17 on Fig. 6-5. This can be accomplished by applying the principles developed in the *Lateral Spillway Channels* section of Chap. 3. In this case the necessary computations can be made by applying Eq. (3-50).

$$y_u = \left[\frac{2(y_c)^3 + (y_l)^3}{y_l} \right]^{1/2}$$

(a) Calculate the critical depth from Eq. (3-18).

$$y_c = \left[\frac{Q^2}{gw^2} \right]^{1/3}$$
$$= \left[\frac{(24.9/2)^2}{(32.2)(4)^2} \right]^{1/3}$$
$$= \textbf{0.67 ft}$$

(b) Compute the depth of flow at point 17 from Eq. (3-50).

$$y_{17} = \left[\frac{2(0.67)^3 + (1.98)^3}{1.98} \right]^{1/2}$$
$$= \textbf{2.06 ft}$$

14. Establish the water surface elevation at point 17.

$$\text{WSE}(17) = \text{CBE}(17) + y_{17}$$
$$= 294.80 + 2.06$$
$$= \textbf{296.86 ft}$$

15. Compare the water surface elevation at point 17 to the elevation of the control weir for the primary clarifier system. This comparison is necessary to ensure that there is at least 2 in. of free-fall from the control weir to the lateral spillway channel.

$$\begin{aligned}
\text{Control weir elevation:}\quad &298.71 \text{ ft} \\
\text{WSE(17):}\quad &296.86 \text{ ft} \\
\text{Difference:}\quad &1.85 \text{ ft}
\end{aligned}$$

This difference is excessive. It should be reduced by raising the elevation of the aeration basin system control weir by an amount equal to 1.85 − 2/12 = **1.68 ft.**

16.   Establish the water surface profile between the aeration basin system and the primary clarifier system based on a 296.98-ft elevation for the aeration basin system control weir and an elevation of 296.98 − 0.5 = 296.48 ft for the aeration basin system distribution channel bottom.

$$
\begin{aligned}
WSE(23) &= 296.26 \text{ ft} + 1.68 \text{ ft} = 297.94 \text{ ft} \\
WSE(22) &= 296.50 \text{ ft} + 1.68 \text{ ft} = 298.18 \text{ ft} \\
WSE(21) &= 296.50 \text{ ft} + 1.68 \text{ ft} = 298.18 \text{ ft} \\
WSE(20) &= 296.62 \text{ ft} + 1.68 \text{ ft} = 298.30 \text{ ft} \\
WSE(19) &= 296.62 \text{ ft} + 1.68 \text{ ft} = 298.30 \text{ ft} \\
WSE(18) &= 296.78 \text{ ft} + 1.68 \text{ ft} = 298.46 \text{ ft} \\
WSE(17) &= 296.82 \text{ ft} + 1.68 \text{ ft} = 298.50 \text{ ft}
\end{aligned}
$$

Aeration Basin Lateral Spillway Channel

The fourth control point for the treatment plant is located at the exit end of the lateral spillway channel for the aeration basin (see Fig. 6-6, point 26). A free overfall is designed to occur at this point, which means critical depth will be established. Hence, the depth of flow at point 25 on Fig. 6-6 may be calculated from Eq. (3-48). The elevation of the bottom of the lateral spillway channel is then established by allowing approximately 2 in. of free overfall from the aeration basin into the lateral spillway channel. In other words, increase the depth of flow at point 25 by approximately 2 in., then subtract this amount from the elevation of the aeration basin control weir. The resulting value is the elevation of the lateral spillway channel bottom. The required computations are described below:

1.   Compute total flow through a single aeration basin.

$$Q = 1/3 \text{ inflow} + 1/3 \text{ sludge return}$$

$$= \frac{24.9}{3} + \frac{(0.55)(24.9)}{3}$$

$$Q = \textbf{12.86 cfs}$$

2.   Calculate the critical depth of flow which assumed to occur at point 26.

$$y_c = \left[\frac{Q^2}{gw^2} \right]^{1/3}$$

$$= \left[\frac{(12.86)^2}{(32.2)(4)^2} \right]^{1/3}$$

$$= \textbf{0.68 ft}$$

3. Determine the depth of flow at point 25.

$$y_{25} = 1.73y_c$$
$$= (1.73)(0.68)$$
$$= \textbf{1.17 ft}$$

4. Establish the elevation of the lateral spillway channel bottom.

$$\begin{bmatrix} \text{Channel bottom} \\ \text{elevation} \end{bmatrix} = \begin{bmatrix} \text{control weir} \\ \text{elevation} \end{bmatrix} - \begin{bmatrix} y_{25} + \dfrac{2}{12} \end{bmatrix}$$

$$= 296.98 - [1.17 + 0.17]$$
$$= \textbf{295.64 ft}$$

6.4 HYDRAULIC ANALYSIS FOR FINAL CLARIFIER SYSTEM

The fifth control point is located at the effluent weir of the final clarifier (see Fig. 6-7). The elevation of this weir will control the water surface elevation in the 2-ft-diameter riser pipe that collects the flow from the lateral spillway channel of an aeration basin. If there were no energy loss due to friction in the piping system from an aeration basin lateral spillway channel to the final clarifier, the water surface elevation in the riser pipe would be the same as the water surface elevation in the final clarifier. However, since there is an energy loss in the piping system, the water surface elevation in the riser will be above the water surface elevation in the clarifier by an amount equal to the energy loss (see Fig. 6-8).

Computations

1. It will be assumed that V-notch weirs will control the flow from the final clarifiers. The weirs will be located around the periphery of the tank and will have 2 notches/ft of length.

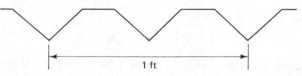

1 ft

2. It will be assumed that the lateral spillway channel can be designed as a straight channel with a length equal to one-half the circumference of the final clarifier and carrying one-half the flow to the clarifier.

3. Determine the number of notches controlling the flow to the lateral spillway channel.

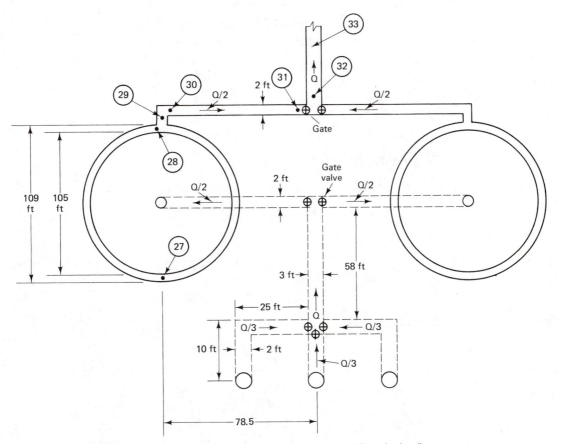

FIGURE 6-7 *Schematic of final clarifier system and ancillary hydraulic components.*

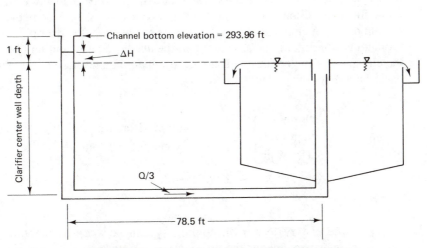

FIGURE 6-8 *Schematic of final clarifier piping system.*

$$\begin{bmatrix} \text{Number of} \\ \text{notches} \end{bmatrix} = \begin{bmatrix} \text{circumference} \\ \text{of clarifier} \end{bmatrix} \Big/ 2 \; (2)$$

$$= \begin{bmatrix} \dfrac{\pi D}{2} \end{bmatrix}(2)$$

$$= \pi(105)$$

$$= 329.8; \text{ use } \textbf{330 notches}$$

4. Compute the flow per V-notch weir.

$$\begin{bmatrix} \text{Flow per} \\ \text{weir} \end{bmatrix} = \dfrac{\begin{bmatrix} \text{flow through} \\ \text{clarifier} \end{bmatrix} \Big/ 2}{330}$$

$$= \begin{bmatrix} \dfrac{(24.9/2)(2)}{330} \end{bmatrix}$$

$$= \textbf{0.019 cfs}$$

5. Calculate the head on the V-notch weir.

$$H = \begin{bmatrix} \dfrac{Q}{2.5} \end{bmatrix}^{0.40}$$

$$H = \begin{bmatrix} \dfrac{0.019}{2.5} \end{bmatrix}^{0.40}$$

$$= \textbf{0.14 ft}$$

6. Establish the elevation of the control weir by arbitrarily assuming a distance below the bottom of the aeration basin lateral spillway channel, which will allow free overfall but will not produce an excessive loss of head between the bottom of the channel and the water surface elevation in the riser pipe. In this problem it will initially be assumed that the above requirement will be met if the elevation of the control weir is 1 ft less than the elevation of the aeration basin lateral spillway channel bottom.

$$\begin{bmatrix} \text{Initial} \\ \text{elevation of} \\ \text{control weir} \end{bmatrix} = \begin{bmatrix} \text{elevation of lateral} \\ \text{spillway channel bottom} \end{bmatrix} - 1 \text{ ft}$$

$$= 295.64 - 1.00$$

$$= \textbf{294.64 ft}$$

7. Compute the energy loss in the piping system between the aeration basin lateral spillway channel and the final clarifier. Assume that uncoated cast-iron

pipe is used throughout. For this system Table 1-5 shows that the Hazen–Williams coefficient, C, has a value of 130.

(a) Calculate the velocity of flow in the riser pipe through the clarifier

$$V = \frac{\left[\dfrac{24.9}{2} + \dfrac{(0.55)(24.9)}{2}\right]}{\pi(2)^2/4}$$

$$= \frac{19.29}{3.14}$$

$$= \textbf{6.14 ft/s}$$

(b) Compute the entrance loss to the clarifier.

$$h_E = K_{ent}\left[\frac{V^2}{2g}\right]$$

where K_{ent} is normally taken as 1.0.

$$= 1.0\left[\frac{(6.14)^2}{2(32.2)}\right]$$

$$= \textbf{0.58 ft}$$

(c) Determine the energy loss due to friction through the clarifier riser pipe.
The slope of the energy line is given by a rearrangement of Eq. (1-29).

$$S = \left[\frac{V}{1.318CR^{0.63}}\right]^{1.85}$$

$$= \left[\frac{6.14}{1.318(130)(1/2)^{0.63}}\right]^{1.85}$$

$$= \textbf{0.0047 ft/ft}$$

The energy loss is, therefore,

$$h_f = (S)(\text{clarifier depth}) + (\text{bottom slope})(\text{radius})$$

$$= (0.0047)(14) + (0.01)(52.5)$$

$$= \textbf{0.07 ft}$$

(d) Calculate the energy loss in the 90° bend at the foot of the riser pipe through the clarifier.

$$h_B = K_{90} \left[\frac{V^2}{2g} \right]$$

where K_{90} is normally taken as 0.3.

$$= 0.3 \left[\frac{(6.14)^2}{2(32.2)} \right]$$

$$= \mathbf{0.17 \ ft}$$

(e) Determine the energy loss due to friction in the pipe feeding the clarifier riser pipe.

$$S = \left[\frac{6.14}{1.318(130)(1/2)^{0.63}} \right]^{1.85}$$

$$= 0.0047 \ ft/ft$$

The energy loss is, therefore,

$$h_f = (S)(\text{pipe length})$$

$$= (0.0047)(78.5)$$

$$= \mathbf{0.37 \ ft}$$

(f) Compute the energy loss in the reducing tee feeding the clarifiers.

$$h_T = K_T \left[\frac{V^2}{2g} \right]$$

where K_T is normally taken as 0.9, based on velocities of the smaller end.

$$h_T = 0.9 \left[\frac{(6.14)^2}{2(32.2)} \right]$$

$$= \mathbf{0.53 \ ft}$$

(g) Calculate the energy loss through the gate valve located at the tee.

$$h_v = K_v \left[\frac{V^2}{2g} \right]$$

Assuming the valve is fully open, K_v is normally taken as 0.19.

$$= 0.19 \left[\frac{(6.14)^2}{2(32.2)} \right]$$

$$= \mathbf{0.11 \ ft}$$

(h) Calculate the velocity of flow in the main line between the aeration basin and the final clarifier.

$$V = \frac{24.9 + (0.55)(24.9)}{\pi(3)^2/4}$$

$$= \textbf{5.46 ft/s}$$

(i) Determine the energy loss due to friction in the main line between the aeration basin and the final clarifier.

$$S = \left[\frac{5.46}{1.318(130)(2.25/3)^{0.63}}\right]^{1.85}$$

$$= 0.0024 \text{ ft/ft}$$

The energy loss is, therefore,

$$h_f = (S)(\text{pipe length})$$

$$= 0.0024 \ (58 \text{ ft})$$

$$= \textbf{0.14 ft}$$

(j) Compute the velocity of flow in the feeder line from the aeration basin.

$$V = \frac{\left[\dfrac{24.9 + (0.55)(24.9)}{3}\right]}{\pi(2)^2/4}$$

$$= \textbf{4.09 ft/s}$$

(k) Compute the energy loss in the reducing tee to the main feeder line.

$$h_T = 0.9 \left[\frac{(4.09)^2}{2(32.2)}\right]$$

$$= \textbf{0.23 ft}$$

(l) Compute the energy loss through the gate valve located at the tee, assuming the valve is fully open.

$$h_v = 0.19 \left[\frac{(4.09)^2}{2(32.2)}\right]$$

$$= \textbf{0.05 ft}$$

(m) Determine the energy loss due to friction in the line from the riser pipe to the aeration basin lateral spillway channel.

$$S = \left[\frac{4.09}{1.318(130)(1/2)^{0.63}}\right]^{1.85}$$

$$= 0.0022 \text{ ft/ft}$$

The energy loss is, therefore,

$$h_f = (S)(\text{pipe length})$$

$$= 0.0022(25 \text{ ft})$$

$$= 0.05 \text{ ft}$$

(n) Calculate the energy loss in the 90° bend at the foot of the riser pipe to the aeration basin lateral spillway channel.

$$h_B = 0.3\left[\frac{(4.09)^2}{2(32.2)}\right]$$

$$= 0.08 \text{ ft}$$

(o) Estimate the energy loss due to friction in the riser pipe to the aeration basin lateral spillway channel.

$$S = \left[\frac{4.09}{1.318(130)(1/2)^{0.63}}\right]^{1.85}$$

$$= 0.0022 \text{ ft/ft}$$

The energy loss is, therefore,

$$h_f = (S)(\text{pipe length})$$

$$= (0.0022)(14 + (0.01)(52.5) + 1.0)$$

$$= 0.03 \text{ ft}$$

(p) Compute the total energy loss in the piping system between the aeration basin lateral spillway channel and the final clarifier.

$$\Delta H = 0.58 + 0.07 + 0.17 + 0.37 + 0.53 + 0.11 + 0.14$$
$$+ 0.23 + 0.05 + 0.05 + 0.08 + 0.03$$
$$= 2.41 \text{ ft}$$

8. Since the total energy loss computed in Step 7(p) is greater than the 1.0 ft allowed for this loss in Step 6, the control weir for the final clarifier must be lowered if free overfall is to occur in the aeration basin lateral spillway channel.

$$\begin{bmatrix} \text{Actual} \\ \text{elevation of} \\ \text{control weir} \end{bmatrix} = \begin{bmatrix} \text{elevation of lateral} \\ \text{spillway channel bottom} \end{bmatrix}$$

$$- \Delta H - H - \text{3-in. free overfall}$$

$$= 295.64 - 2.41 - 0.14 - 0.25$$

$$= \mathbf{292.84 \ ft}$$

6.5 HYDRAULIC ANALYSIS FOR CHLORINE CONTACT BASIN

The sixth control point for the treatment plant is located at the effluent weir for the chlorine contact basin. The elevation between point 35 on Fig. 6-9 and point 27 on Fig. 6-7, i.e., between the fifth and sixth control points. To make the water surface elevation calculations between these two control points, it is first necessary to establish the control weir elevation for the chlorine contact basin. Once this value is set, the depth of flow at point 28 can be established by making calculations similar to those carried out for the bar screen channel. After the depth of flow at point 28 is computed, the depth of flow at point 27 is calculated from Eq. (3-52). The water surface elevation at point 27 is then determined and compared to the control weir elevation for the final clarifier. The water surface elevation at point 27 should be at least 2 in. below the control weir elevation.

Computations

1. As a first approximation assume that the control weir elevation for the chlorine contact basin is 1.8 ft less than the elevation of the control weir for

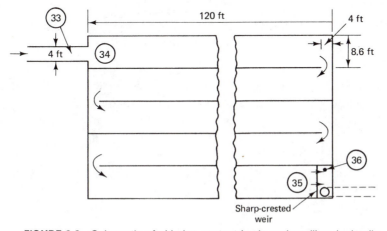

FIGURE 6-9 Schematic of chlorine contact basin and ancillary hydraulic components.

the final clarifier, i.e., the elevation of the chlorine contact basin control weir is 292.84 ft − 1.80 ft = 291.04 ft.

2. Calculate the head on the control weir from Eq. (1-35).

$$H = \left[\frac{1}{(3.33)^{2/3}(8.6)^{2/3}} \right] (24.9)^{2/3}$$

$$= \textbf{0.90 ft}$$

3. Estimate the energy loss through the chlorine contact basin. For $(n-1)$ equally spaced *over-and-under* or *around-the-end* baffles, Fair et al. (1968) indicated that the loss of head can be approximated from the relationship

$$h_L = n \frac{V_1^2}{2g} + (n-1) \frac{V_2^2}{2g}$$

where V_1 represents the average velocity of flow in the channels and V_2 represents the average velocity of flow in the baffle slots. Hence, for the chlorine contact basin shown in Fig. 6-9

$$h_L = (5) \left[\frac{[(24.9)/(8.4)(8.6)]^2}{2(32.2)} \right] + (4) \left[\frac{[(24.9)/(8.4)(4)]^2}{2(32.2)} \right].$$

$$= 0.009 + 0.034$$

$$= \textbf{0.04 ft}$$

4. Establish the water surface elevation at point 34 shown on Fig. 6-9.

$$\text{WSE}(34) = \left[\begin{array}{c} \text{control weir} \\ \text{elevation} \end{array} \right] + H + h_L$$

$$= 291.04 + 0.09 + 0.04$$

$$= \textbf{291.98 ft}$$

5. Assume a desired depth of flow in the feed channel to the chlorine contact basin. For a channel width of 4.0 ft and a flow of 24.9 cfs, the calculations presented in Step 4 of Sec. 6.1 show that a flow depth greater than 1.064 ft is necessary to prevent a choke condition. A value of $y_{33} = 1.2$ ft will, therefore, be assumed.

6. Apply the energy equation between points 33 and 34 and establish the water surface elevation at point 33.

$$Z_{33} + \frac{V_{33}^2}{2g} = Z_{34} + \frac{V_{34}^2}{2g} + K_{\text{ent}} \frac{V_{33}^2}{2g}$$

By neglecting the velocity head in the chlorine contact chamber and letting the entrance loss coefficient equal 1.0, the following relationship is obtained:

$$Z_{33} = Z_{34}$$

or

$$WSE(33) = WSE(34)$$
$$= 291.98 \text{ ft}$$

7. Establish the channel bottom elevation at point 33.

$$CBE(33) = WSE(33) + y_{33}$$
$$= 291.98 + 1.2$$
$$= \mathbf{290.78 \text{ ft}}$$

8. Assuming no energy loss due to friction, establish the water surface elevation and the depth of flow at point 32 shown on Fig. 6-7.

$$WSE(32) = WSE(33)$$
$$= 291.98$$
$$y_{32} = y_{33}$$
$$= \mathbf{1.2}$$

9. Apply the energy equation between points 31 and 32 shown on Fig. 6-7 and establish the depth of flow at point 31.

$$y_{31} + \frac{V_{31}^2}{2g} = y_{32} + \frac{V_{32}^2}{2g} + K_{gate}\frac{V_{31}^2}{2g} + K_{90}\frac{V_{31}^2}{2g}$$

or

$$y_{31} + (1 - K_{90} - K_{gate})\frac{Q_{31}^2}{(64.4)w_{31}^2 y_{31}^2} = y_{32} + \frac{Q_{32}^2}{(64.4)w_{32}^2 y_{32}^2}$$

Assuming a value of 0.3 for K_{90} and 0.2 for K_{gate}

$$y_{31} + (1 - 0.3 - 0.2)\frac{(24.9/2)^2}{(64.4)(2)^2 y_{31}^2} = 1.2 + \frac{(24.9)^2}{(64.4)(4)^2(1.2)^2}$$

$$y_{31} + \frac{0.30}{y_{31}^2} = 1.62$$

The solution to this equation is found to be $y_{31} = $ **1.48 ft.**

10. Establish the water surface elevation at point 31.

$$WSE(31) = CBE(31) + y_{31}$$
$$= 290.78 + 1.48$$
$$= \mathbf{292.26\ ft}$$

11. Assuming no energy loss due to friction, establish the water surface elevation and the depth of flow at point 30 shown on Fig. 6-7.

$$WSE(30) = WSE(31)$$
$$= 292.26\ ft$$

$$y_{30} = y_{31}$$
$$= \mathbf{1.48\ ft}$$

12. Apply the energy equation between points 29 and 30 shown on Fig. 6-7 and establish the depth of flow at point 29.

$$y_{29} + \frac{V_{29}^2}{2g} = y_{30} + \frac{V_{30}^2}{2g} + K_{90}\frac{V_{29}^2}{2g}$$

or

$$y_{29} + (1 - K_{90})\frac{Q_{29}^2}{(64.4)w_{29}^2 y_{29}^2} = y_{30} + \frac{Q_{30}^2}{(64.4)w_{30}^2 y_{30}^2}$$

Assuming a value of 0.3 for K_{90}

$$y_{29} + (1 - 0.3)\frac{(24.9/2)^2}{(64.4)(2)^2 y_{29}^2} = 1.48 + \frac{(24.9/2)^2}{(64.4)(2)^2(1.48)^2}$$

$$y_{29} + \frac{0.42}{y_{29}^2} = 1.75$$

The solution to this equation is found to be $y_{29} = $ **1.58 ft.**

13. Establish the water surface elevation at point 29.

$$WSE(29) = CBE(29) + y_{29}$$
$$= 290.78 + 1.58$$
$$= \mathbf{292.36\ ft}$$

14. Apply the energy equation between points 28 and 29 shown on Fig. 6-7 and establish the depth of flow at point 28.

$$y_{28} + \frac{V_{28}^2}{2g} = y_{29} + \frac{V_{29}^2}{2g} + K_{90}\frac{V_{28}^2}{2g}$$

or

$$y_{28} + (1 - K_{90})\frac{Q_{28}^2}{(64.4)w_{28}^2 y_{28}^2} = y_{29} + \frac{Q_{29}^2}{(64.4)w_{29}^2 y_{29}^2}$$

Assuming a value of 0.3 for K_{90}

$$y_{28} + (1 - 0.3)\frac{(24.9/4)^2}{(64.4)(2)^2 y_{28}^2} = 1.58 + \frac{(24.9/2)^2}{(64.4)(2)^2(1.58)^2}$$

$$y_{28} + \frac{0.11}{y_{28}^2} = 1.82$$

The solution to this equation is found to be $y_{28} = $ **1.79 ft.**

15. Establish the water surface elevation at point 28.

$$\text{WSE}(28) = \text{CBE}(28) + y_{28}$$
$$= 290.78 + 1.79$$
$$= \textbf{292.57 ft}$$

16. Determine the depth of flow at point 27 on Fig. 6-7 by applying Eq. (3-52).

(a) Compute the depth of flow at point 27 from Eq. (3-52).

$$y_{27} = \left[\frac{2(mq)^2}{gw^2 y_{28}} + y_{28}^2\right]^{1/2}$$
$$= \left[\frac{2[(330)(0.019)]^2}{(32.2)(2)^2(1.79)} + (1.79)^2\right]^{1/2}$$
$$y_{27} = \textbf{1.88 ft}$$

(b) Establish the water surface elevation at point 27.

$$\text{WSE}(27) = \text{CBE}(27) + y_{27}$$
$$= 290.78 + 1.88$$
$$= \textbf{292.66 ft}$$

17. Compare the water surface elevation at point 27 to the elevation of the control weir for the final clarifier system. This comparison is necessary to ensure that there are at least 2 in. of free-fall from the control weir to the lateral spillway channel.

Control weir elevation: 292.84 ft

WSE(27):   292.66 ft

Difference: 0.18 ft

This difference is slightly large, but it will be accepted for this problem. It could easily be reduced by raising the chlorine contact basin control weir.

The hydraulic analysis will be terminated at this point. The water surface elevation between point 36 shown on Fig. 6-9 and the water surface elevation in the receiving stream can be established by the same basic principles applied throughout this chapter and presented in Chaps. 1 through 4.

REFERENCES

FAIR, G. M., GEYER, J. C., and OKUN, D. A., *Water and Wastewater Engineering: Volume 2,* John Wiley & Sons, New York, New York (1968).

METCALF and EDDY, INC., *Wastewater Engineering: Treatment Disposal, Reuse,* 2nd Edition, McGraw–Hill Book Company, New York, New York (1979).

APPENDICES

APPENDIX I: FORTRAN COMPUTER CODES FOR SELECTED PROBLEMS

Computer Code for Example Problem 1-9

```
      C              AM=CROSS SECTIONAL AREA OF MANIFOLD
      C              AL=CROSS SECTIONAL AREA OF LATERAL
      C              QL=FLOW IN LATERAL
      C              VL=VELOCITY IN LATERAL
      C              QO=INITIAL FLOW
      C              QM=FLOW IN MANIFOLD UPSTREAM OF LATERAL
      C              VM=VELOCITY IN MANIFOLD UPSTREAM OF LATERAL
      C              B=BETA TERM
      C              N=NUMBER OF LATERALS
    1              DIMENSION QL(30),VL(30),QM(30),VM(30),B(30),CQL(30)
    2              SB=0.
    3              QO=15.47
    4              AM=0.785
    5              AL=0.087
    6              PHI=1.67
    7              THETA=0.7
    8              N=8
      C              INITIALIZE FLOW DISTRIBUTION
    9              QMI=QO/FLOAT(N)
   10              DO 10 I=1,N,1
   11              QL(I)=QMI
   12              VL(I)=QL(I)/AL
   13              CQL(I)=QL(I)
   14        10    CONTINUE
   15              DO 20 I=1,N,1
   16              IF(I.EQ.1) GO TO 15
   17              QM(I)=QM(I-1)-QL(I-1)
   18              GO TO 18
   19        15    QM(I)=QO
   20        18    VM(I)=QM(I)/AM
   21        20    CONTINUE
   22              DO 30 I=1,N,1
   23              VR=(VM(I)/VL(I))**2
   24              B(I)=(PHI*VR)+THETA+1.0
```

```
25                    RB=SQRT(1./B(I))
26                    SB=RB+SB
27           30       CONTINUE
28           35       VL(1)=(QO/(AL*SQRT(B(1))))*(1./SB)
29                    DO 40 I=2,N,1
30                    VL(I)=VL(1)*SQRT(B(1)/B(I))
31           40       CONTINUE
32                    DO 50 I=1,N,1
33                    QL(I)=AL*VL(I)
34           50       CONTINUE
35                    DO 60 I=1,N,1
36                    IF(I.EQ.1) GO TO 65
37                    QM(I)=QM(I-1)-QL(I-1)
38                    GO TO 68
39           65       QM(I)=QO
40           68       VM(I)=QM(I)/AM
41           60       CONTINUE
42                    SQ=0.
43                    SB=0.
44                    DO 70 I=1,N,1
45                    VR=(VM(I)/VL(I))**2
46                    B(I)=(PHI*VR)+THETA+1.0
47                    RB=SQRT(1./B(I))
48                    SB=RB+SB
49                    SQ=QL(I)+SQ
50           70       CONTINUE
51                    DO 80 I=1,N,1
52                    WRITE(6,90)QL(I),VL(I),QM(I),VM(I)
53           90       FORMAT('0',1X,9HQL EQUALS, 1X,F8.3,5X,9HVL EQUALS, 1X,F8.3,5X,
                     $9HQM EQUALS,1X,F8.3,5X,9HVM EQUALS,1X,F8.3)
54           80       CONTINUE
55                    WRITE(6,100) SQ
56           100      FORMAT('0',1X,9HSQ EQUALS,1X,F8.3)
57                    WRITE(6,105)
58           105      FORMAT('0',1X,130('*'))
59                    NN=0
60                    DO 110 I=1,N,1
61                    IF(ABS((CQL(I)-QL(I))/QL(I)).LT.1.E-2) NN=NN+1
62           110      CONTINUE
63                    IF(NN.EQ.N) GO TO 115
64                    DO 120 I=1,N,1
65                    CQL(I)=QL(I)
66           120      CONTINUE
67                    GO TO 35
68           115      STOP
69                    END

      /GO
```

General Program for Example Problems 2-1, 2-2 and 2-3

```
C  THIS PROGRAM WAS RUN ON A TSS.  MINOR LOSSES WERE NEGLECTED.
C  NPRT(I)    = NUMBER OF PORTS IN ITH SECTION
C  DPORT(I)   = PORT DIAMETER FOR ITH SECTION
C  DPIPE(I)   = PIPE DIAMETER FOR ITH SECTION
C  SPAC(I)    = PORT SPACING FOR ITH SECTION
C  F(1)       = FRICTION FACTOR FOR ITH SECTION
C  QSEC(I)    = TOTAL DISCHARGE FROM ITH SECTION
C  E(J)       = ENERGY AT JTH PORT
C  Q(J)       = DISCHARGE FROM JTH PORT
C  V(J)       = PIPE VELOCITY UPSTREAM FROM JTH PORT
C  HF(J)      = FRICTIONAL HEAD LOSS BETWEEN JTH AND (J+1) PORT
C  CD(J)      = DISCHARGE COEFFICIENT OF JTH PORT
C  QN(J)      = QAVE/QN(J) = NORMALIZED PORT DISCHARGE
C  M          = NUMBER OF DIFFUSER SECTIONS
```

```
C   N           = TOTAL NUMBER OF DIFFUSER PORTS
C   QAVE        = AVERAGE PORT DISCHARGE FOR DIFFUSER
C   G           = GRAVITATIONAL ACCELERATION
C   GAMMAR      = SPECIFIC WEIGHT OF RECEIVING WATER
C   GAMMAE      = SPECIFIC WEIGHT OF EFFLUENT
C   SLOPE       = SLOPE OF DIFFUSER PIPE (ASSUMED CONSTANT)
C   DSCHG       = TOTAL DIFFUSER DISCHARGE
C   E(1)        = ENERGY AT DOWNSTREAM PORT
        G=32.2
        PRINT 736
    736 FORMAT (/'SPECIFIC WEIGHT OF EFFLUENT')
        READ,GAMMAE
        PRINT 731
    731 FORMAT (/'SPECIFIC WEIGHT OF RECEIVING WATER')
        READ,GAMMAR
        IF (GAMMAR.EQ.GAMMAE) GO TO 732
        PRINT 733
    733 FORMAT (/'SLOPE OF DIFFUSER PIPE')
        READ,SLOPE
    732 CONTINUE
        DIMENSION NPRT(10),DPORT(10),DPIPE(10),SPAC(10),F(10),QSEC(10)
        DIMENSION E(200),Q(200),V(200),HF(200),CD(200),QN(200)
        N=0
        PRINT 700
    700 FORMAT (/'NUMBER OF DIFFUSER SECTIONS')
        READ,M
        DO 701 I = 1, M
        PRINT 702,I
    702 FORMAT ('NUMBER OF PORTS IN SECTION',I2)
        READ,NPRT(I)
        PRINT 703,I
    703 FORMAT ('DIAMETER OF PORTS FOR SECTION',I2,' IN FEET')
        READ,DPORT(I)
        PRINT 704,I
    704 FORMAT ('DIAMETER OF DIFFUSER PIPE FOR SECTION',I2,'IN FEET')
        READ,DPIPE(I)
        PRINT 705,I
    705 FORMAT ('PORT SPACING FOR SECTION',I2,' IN FEET')
        READ,SPAC(I)
        PRINT 706,I
    706 FORMAT ('FRICTION FACTOR, F, FOR SECTION',I2)
        READ,F(I)
    701 N = N+NPRT(I)
        PRINT 720
    720 FORMAT (/'INITIAL ESTIMATE OF DOWNSTREAM ENERGY, E(1), IN FEET')
        READ,E(1)
    719 R = 0
        DSCHG = 0.
        DO 718 I = 1,M
        QSEC(I) = 0.
        DO 718 J = 1,NPRT(I)
        K = K + 1
        IF (I+J-2) 707,707,708
C   THE DISCHARGE COEFFICIENT FOR SHARP-EDGED ORIFICES IS USED HEREIN.
    707 CD(K) = 0.63
        GO TO 709
    708 CD(K) = 0.63-0.58*V(K-1)**2./(2.*G*E(K))
    709 Q(K) = CD(K)*(3.1416/4.)*DPORT(I)**2.*(2.*G*E(K))**.5
        DSCHG = DSCHG+Q(K)
        V(K) = 4.*DSCHG/(3.1416*DPIPE(I)**2)
    712 HF(K) = (F(I)*SPAC(I)/DPIPE(I))*V(K)**2/(2.*G)
        QSEC(I) = QSEC(I) + Q(K)
    718 E(K+1) = E(K)+HF(K)+(GAMMAR-GAMMAE)*SLOPE*SPAC(I)/GAMMAE
        PRINT 714,E(N),DSCHG
    714 FORMAT (/'ENERGY UPSTREAM = ',F5.3,' AND TOTAL DISCHARGE = ',F6.3)
        PRINT 722
```

```
          722 FORMAT (/'OUTPUT FOR EACH PORT?   (YES=1,NO=0)')
              READ,OUTPUT
              IF (OUTPUT.NE.0) GO TO 723
          727 PRINT 716
          716 FORMAT (/'NEW ESTIMATE OF DOWNSTREAM ENERGY, E(1), IN FEET')
    C ENTER ZERO FOR E(1) IF RESULTS ARE SATISFACTORY; OTHERWISE, ENTER NEW E(1)
              READ,E(1)
              IF (E(1).NE.0.) GO TO 719
              GO TO 728
          723 PRINT 724
          724 FORMAT (/'PORT    E       Q      V      HF     CE     QN')
              QAVE = DSCHG/N
              DO 726 K = 1,N
              QN(K) = W(K)/QAVE
              PRINT 725,K,E(K),N(K),V(K),HF(K),CD(K),QN(K)
          725 FORMAT (1X,I2,2X,6F7.3)
          726 CONTINUE
              DO 735 I = 1,M
              PRINT 734,I,QSEC(I)
          734 FORMAT ('DISCHARGE IN SECTION',I2,' = ',F6.3)
          735 CONTINUE
              GO TO 727
          728 STOP
              END
```

Computer Code for Example Problem 3-3

```
1                  Q=12.
2                  FN=0.013
3                  W=2.5
4                  S=0.001
    C  COMPUTE NORMAL DEPTH
5                  YO=0.1
6            10    Q1=(W*YO)*(1.486/FN)*(((W*YO)/((2.*YO)+W))**0.67)*(S**0.5)
7                  IF(ABS((Q1-Q)/Q).LT.3.2E-2) GO TO 20
8                  YO=YO + 0.1
9                  WRITE(6,9) YO
10           9     FORMAT('0',1X,9HYO EQUALS,1X,F10.4)
11                 IF(YO.GT.4.) GO TO 50
12                 GO TO 10
    C  COMPUTE CRITICAL DEPTH
13           20    YC=((Q**2)/(32.2*(W**2)))**0.33
    C  COMPUTE DEPTH AT BRINK
14                 YB=0.72*YC
15                 WRITE(6,21) YO
16           21    FORMAT('0',1X,19HNORMAL DEPTH EQUALS,1X,F10.4)
17                 WRITE(6,22) YO
18           22    FORMAT('0',1X,21HCRITICAL DEPTH EQUALS,1X,F10.4)
19                 WRITE(6,23) YE
20           23    FORMAT('0',1X,18HBRINK DEPTH EQUALS,1X,F10.4)
21                 WRITE(6,24)
22           24    FORMAT('0',13X,2HY1,13X,2HY2,13X,2HV1,13X2HV2,13X,4HRAVE
                   $13X4HSAVE,13X,2HDX,13X,3HCDX)
    C  SET UP 0.1 FT INCREMENTS
23                 Y2=YC
24                 CDX=0.0
25           30    V2=Q/(W*Y2)
26                 Y1=Y2+.1
27                 VI=Q/(W*Y1)
28                 R1=(W*Y1)/((2.*Y1)+W)
29                 R2=(W*Y2)/((2.*Y2)+W)
30                 RAVE=(R1+R2)/2.0
31                 SAVE=((FN**2)*((VI+V2)**2))/(8.83*(RAVE**1.33))
32                 E1=Y1+((VI**2)/64/4)
33                 E2=Y2+((V2**2)/64.4)
34                 DX=(E1-E2)/(SAVE-S)
```

```
35                  CDX=CDX+DX
36                  WRITE(6,31) Y1,Y2,V1,V2,RAVE,SAVE,DX,CDX
37          31      FORMAT('0',5X,F10.5,5X,F10.5,5X,F10.5,5X,F10.5,5X,F10.5,
                    $F10.5,5X,F10.5,5X,F10.5)
38                  IF((YO-Y1).LT.0.0) GO TO 32
39                  Y2=Y1
40                  GO TO 30
41          32      Y1=YO
42                  CDX=0.0
43          40      V1=Q/(W*Y1)
44                  Y2=Y1-0.1
45                  V2=Q/(W*Y2)
46                  R1=(W*Y1)/((2.*Y1)+W)
47                  R2=(W*Y2)/((2.*Y2)+W)
48                  RAVE=(R1+R2)/2.0
49                  SAVE=((FN**2)*((V1+V2)**2))/(8.83*(RAVE**1.33))
50                  E1=Y1+((V1**2)/64.4)
51                  E2=Y2+((V2**2)/64.4)
52                  DX=(E1-E2)/(SAVE-S)
53                  CDX=CDX+DX
54                  WRITE(6,41) Y1, Y2,V1,V2,RAVE,SAVE,DX,CDX
55          41      FORMAT('0',5X,F10.5,5X,F10.5,5X,F10.5,5X,F10.55X,F10.55X,
                    $F10.5,5X,F10.5,5X,F10.5)
56                  IF((Y2-YB).LT.0.0) GO TO 50
57                  Y1=Y2
58                  GO TO 40
59          50      STOP
60                  END

            /GO
```

Computer Code for Example Problem 3-4

```
1                   FN=.013
2                   W=1.5
3                   Q=6.1
4                   DX=1.0
5                   RL=24.
6                   S=.001
7                   G=32.2
8                   XS=0.0
9                   WRITE(6,16)
10          16      FORMAT('0',10X,2HY1,10X,2HY2,10X,2HQ1,10X,2HQ2,10X,2HV1,
                    $10X,2HV2,10X,2HX1,10X,2HX2,10X,4HDYPA,10X,4HDYPC)
11                  QS=Q/RL
12                  YC=((Q**2)/(G*(W**2)))**.333
13                  X2=RL
14                  Y2=YC
15          24      DYPA=0.0005
16          25      DY=(S*DX)-DYPA
17                  Y1=Y2-DY
18                  Q2=QS*X2
19                  Q1=QS*(X2-DX)
20                  V2=Q2/(W*Y2)
21                  V1=Q1/(W*Y1)
22                  R1=(W*Y1)/((2.*Y1)+W)
23                  R2=(W*Y2)/((2.*Y2)+W)
24                  RAVE=(R1+R2)/2.
25                  SAVE=((FN**2)*((V1+V2)**2))/(8.83*(RAVE**1.33))
26                  CC1=(Q1*(V1+V2))/(G*(Q1+Q2))
27                  DV=V2-V1
28                  DQ=Q2-Q1
29                  CC2=DV+((V2/Q1)*DQ)
30                  DYPC=(CC1*CC2)+(SAVE*DX)
31                  IF(ABS((DYPA-DYPC)/DYPC).LT.1.0E-1) GO TO 30
32                  IF(DYPA.GT.10.) GO TO 79
33                  DYPA=DYPA+0.0005
34                  GO TO 25
```

```
35          30    X1=X2-DX
36                WRITE(6,31) Y1,Y2,Q1,Q2,V1,V2,X1,X2,DYPA,DYPC
37          31    FORMAT('0',1X,10F10.4)
38                IF(X1.LT.DX) GO TO 40
39                XS=XS+DX
40                X2=X1
41                Y2=Y1
42                GO TO 24
43          40    DYPA=(2.*(V1**2))/(2.*G)
44                DYPC=DYPA
45                X2=X1
46                Y2=Y1
47                V2=V1
48                Q2=Q1
49                V1=0.0
50                DXF=X-XS
51                DY=(S*DXF)-DYPA
52                Y1=Y2-DY
53                X1=0.0
54                WRITE(6,41) Y1,Y2,Q1,Q2,V1,V2,X1,X2,DYPA,DYPC
55          41    FORMAT('0',1X,10F10.4)
56                GO TO 100
57          79    WRITE(6,80)
58          80    FORMAT('0',1X,20HDYPA GREATER THAN 10)
59         100    STOP
60                END

           /GO
```

Computer Code for Example Problem 3-7

```
1                 DIMENSION QWU(6),YWU(6),FU(6), QW(6),YWD(6),QU(6),YWA(6),
                 $QWC(6),VU(6)
      C  QD=TOTAL FLOW
      C  WL=WEIR LENGTH
      C  W=WIDTH CHANNEL
      C  N=NUMBER OF WEIRS
      C  B=WEIR HEIGHT ABOVE FLOOR
      C  G=GRAVITY CONSTANT
2                 QD=23.21
3                 WL=4.
4                 W=4.
5                 N=6
6                 B=2.0
7                 G=32.2
8                 QW(N)=QD/FLOAT(N)
9          10     YWU(N)=B+0.01
10         15     VU(N)=QW(N)/(W*YWU(N))
11                FU(N)=VU(N)/((G*YWU(N))**.5)
12                A1=3.*(FU(N)**2)
13                A2=(FU(N)**2)+2.
14                A3=A1/A2
15                CW=0.611*((1.-A3)**.5)
16                QWC(N)=0.67*W1*CW*((2.*G)**.5)*((YWU(N)-B)**1.5)
17                DQW=ABS(QWC(N)-QW(N))
18                IF(DQW.GT.0.01) GO TO 20
19                I=0
20                E=YWU(N)+(((QWC(N)/(W*YWU(N)))**2)/(2.*G))
21                R=QWC(N)
22                GO TO 30
23         20     IF(QWC(N).GT.QW(N)) GO TO 21
24                IF(QWC(N).LT.QW(N)) GO TO 22
25         21     YWU(N)=YWU(N)-0.005
26                GO TO 15
27         22     YWU(N)=YWU(N)+0.005
```

```
28                        IF(YWU(N).GT.8.) GO TO 91
29                        GO TO 15
30              30        I=I+1
31                        K=N-I
32                        QW(K)=QWC(K+1)
33                        YWD(K)=YWU(K+1)
34              32        QU(K)=QW(K)+R
35                        YWU(K)=B+0.005
36              35        B1=W*YWU(K)
37                        IF(E-YWU(K).LT.0.001) GO TO 40
38                        B2=(2.*G*(E-YWU(K)))**.5
39                        B6=B1*B2
40                        IF(ABS((QU(K)-B6)/QU(K)).LT.1.0E-1) GO TO 40
41                        YWU(K)=YWU(K)+0.005
42                        IF(YWU(K).GT.8.) GO TO 91
43                        GO TO 35
44              40        YWA(K)=(YWD(K)+YWU(K))/2.
45                        VU(K)=QU(K)/(W*YWU(K))
46                        FU(K)=VU(K)/((G*YWU(K))**.5)
47                        A1=3.*(FU(K)**2)
48                        A2=(FU(K)**2)+2.
49                        A3=A1/A2
50                        CW=0.611*((1.-A3)**.5)
51                        QWC(K)=0.67*WL*CW*((2.*G)**.5)*((YWA(K)-B)**1.5)
52                        DQW=ABS(QWC(K)-QW(K))
53                        IF(DQW.GT.0.01) GO TO 50
54                        WRITE(6,31) QW(N),QWC(K),R,CW,YWU(K),YWD(K),FU(K),
                        $VU(K),K
55              31        FORMAT('0',5X,8F10.4,5X,12)
56                        GO TO 60
57              50        IF(QWC(K).LT.QW(K)) GO TO 51
58                        IF(QWC(K).GT.QW(K)) GO TO 52
59              51        QW(K)=QW(K)-0.005
60                        GO TO 32
61              52        QW(K)=QW(K)+0.005
62                        IF(QW(K).GT.(2.*QD)) GO TO 93
63                        GO TO 32
64              60        IF(K.EQ.1) GO TO 70
65                        R=R+QWC(K)
66                        GO TO 30
67              70        QT=QWC(1)+R
68                        WRITE(6,71) QT
69              71        FORMAT('0',20X,F10.4)
70                        DQ=ABS(QD-QT)
71                        IF(QD.LT.(FLOAT(N)*0.05)) GO TO 80
72                        IF(QD.LT.QT) GO TO 75
73                        IF(QD.GT.QT) GO TO 76
74              75        QW(N)=QW(N)-0.05
75                        GO TO 10
76              76        QW(N)=QW(N)+0.1
77                        GO TO 10
78              80        DO 90 J=1,N,1
79                        WRITE(6,81) J,QWC(J),YWU(J)
80              81        FORMAT('0',1X11HWEIR NUMBER,1X,12,10X,11HFLOW EQUALS,
                        $1X,F10.4,10X,21HUPSTREAM DEPTH EQUALS,1X,F10.4)
81              90        CONTINUE
82                        GO TO 100
83              91        WRITE(6,92)
84              92        FORMAT('0',1X,17HYW GREATER THAN 8)
85                        GO TO 100
86              93        WRITE(6,94)
87              94        FORMAT('0',1X,19HQW GREATER THAN 2QD)
88              100       STOP
89                        END

        /GO
```

Computer Code for Example Problem 3-7 (Varying Weir Elevations)

```
1                      DIMENSION QWU(6), YWU(6),FU(6),QW(6),YWD(6),QU(6),
                    $YWA(6),QWC(6),VU(6),B(6)
          C  QD=TOTAL FLOW
          C  WS=WEIR LENGTH
          C  W=WIDTH CHANNEL
          C  N=NUMBER OF WEIRS
          C  B=WEIR HEIGHT ABOVE FLOOR
          C  G=GRAVITY CONSTANT
2                      QD=23.21
3                      WL=4.
4                      W=4.
5                      N=6
6                      B(1)=1.90
7                      B(2)=1.94
8                      B(3)=1.96
9                      B(4)=1.98
10                     B(5)=1.99
11                     B(6)=2.00
12                     G=32.2
13                     QW(N)=QD/FLOAT(N)
14            10       YWU(N)=B(N)+0.01
15            15       VU(N)=QW(N)/(W*YWU(N))
16                     FU(N)=VU(N)/((G*YWU(N))**.5)
17                     A1=3.*(FU(N)**2)
18                     A2=(FU(N)**2)+2.
19                     A3=A1/A2
20                     CW=0.611*((1.-A3)**.5)
21                     QWC(N)=0.67*WL*CW*((2.*G)**.5)*((YWU(N)-B(N))**1.5)
22                     DQW=ABS(QWC(N)-QW(N))
23                     IF(DQW.GT.0.01) GO TO 20
24                     I=0
25                     E=YWU(N)+(((QWC(N)/(W*YWU(N)))**2)/(2.*G))
26                     R=QWC(N)
27                     TO TO 30
28            20       IF(QWC(N).GT.QW(N)) GO TO 21
29                     IF(QWC(N).LT.QW(N)) GO TO 22
30            21       YWU(N)=YWU(N)-0.005
31                     GO TO 15
32            22       YWU(N)=YWU(N)+0.005
33                     IF(YWU(N).GT.8.) GO TO 91
34                     GO TO 15
35            30       I=I+1
36                     K=N-I
37                     QW(K)=QWC(K+1)
38                     YWD(K)=YWU(K+1)
39            32       QU(K)=QW(K)+R
40                     YWU(K)=B(K)+0.005
41            35       B1=W*YWU(K)
42                     IF(E-YWU(K).LT.0.001) GO TO 40
43                     B2=(2.*G*(E-YWU(K)))**.5
44                     B6=B1*B2
45                     IF(ABS((QU(K)-B6)/QU(K)).LT.1.0E-1) GO TO 40
46                     YWU(K)=YWU(K)+0.005
47                     IF(YWU(K).GT.8.) GO TO 91
48                     GO TO 35
49            40       YWA(K)=(YWD(K)+YWU(K))/2.
50                     VU(K)=QU(K)/(W*YWU(K))
51                     FU(K)=VU(K)/((G*YWU(K))**.5)
52                     A1=3.*(FU(K)**2)
53                     A2=(FU(K)**2)+2.
54                     A3=A1/A2
55                     CW=0.611*((1.-A3)**.5)
56                     QWC(K)=0.67*WL*CW*((2.*G)**.5)((YWA(K)-B(K))**1.5)
```

```
57                    DQW=ABS(QWC(K)-QW(K))
58                    IF(DQW.GT.0.01) GO TO 50
59                    WRITE(6,31) QW(N),QWC(K),R,CW,YWU(K),YWD(K),FU(K),
                      $VU(K),K
60           31       FORMAT('0',5X,8F10.4,5X,I2)
61                    GO TO 60
62           50       IF(QWC(K).LT.QW(K)) GO TO 51
63                    IF(QWC(K).GT.QW(K)) GO TO 52
64           51       QW(K)=QW(K)-0.005
65                    GO TO 32
66           52       QW(K)=QW(K)+0.005
67                    IF(QW(K).GT.(2.*QD)) GO TO 93
68                    GO TO 32
69           60       IF(K.EQ.1) GO TO 70
70                    R=R+QWC(K)
71                    GO TO 30
72           70       QT=QWC(1)+R
73                    WRITE(6,71) QT
74           71       FORMAT('0',20X,F10.4)
75                    DQ=ABS(QD-QT)
76                    IF(QD.LT.(FLOAT(N)*0.05)) GO TO 80
77                    IF(QD.LT.QT) GO TO 75
78                    IF(QD.GT.QT) GO TO 76
79           75       QW(N)=QW(N)-0.01
80                    GO TO 10
81           76       QW(N)=QW(N)+0.05
82                    GO TO 10
83           80       DO 90 J=1,N,1
84                    WRITE(6,81) J,QWC(J),YWU(J)
85           81       FORMAT('0',1X,11HWEIR NUMBER,1X,I2,10X,11HFLOW EQUALS,
                      $1X,F10.4,10X21HUPSTREAM DEPTH EQUALS,1X,F10.4)
86           90       CONTINUE
87                    GO TO 100
88           91       WRITE(6,92)
89           92       FORMAT('0',1X,17HYW GREATER THAN 8)
90                    GO TO 100
91           93       WRITE(6,94)
92           94       FORMAT('0',1X,19HQW GREATER THAN 2QD)
93           100      STOP
94                    END

        /GO
```

Computer Code for Example Problem 3-7 (Tapered Channel)

```
1                     DIMENSION QWU(6),YWU(6),FU(6),QW(6),YWD(6),QU(6),YWA(6),
                      $QWC(6),VUC(6),WU(6),WD(6)
      C  QD=TOTAL FLOW
      C  WL=WEIR LENGTH
      C  W=WIDTH CHANNEL
      C  N=NUMBER OF WEIRS
      C  B=WEIR HEIGHT ABOVE FLOOR
      C  G=GRAVITY CONSTANT
2                     QD=23.21
3                     WL=4.
4                     N=6
5                     B=2.0
6                     G=32.2
7                     UW=4.
8                     RL=48.
9                     CL=6.
10                    QW(N)=QD/FLOAT(N)
11           10       YWU(N)=B+0.01
12                    WU(N)=(UW/RL)*CL+2.
```

```
13        15    VU(N)=QW(N)/(WU(N)*YWU(N))
14              FU(N)=VU(N)/((G*YWU(N))**.5)
15              A1=3.*(FU(N)**2)
16              A2=(FU(N)**2)+2.
17              A3=A1/A2
18              CW=0.611*((1.-A3)**.5)
19              QWC(N)=0.67*WL*CW*((2.*G)**.5)*((YWU(N)-B)**1.5)
20              DQW=ABS(QWC(N)-QW(N))
21              IF(DQW.GT.0.01) GO TO 20
22              I=0
23              E=YWU(N)+(((QWC(N)/(WU(N)*YWU(N)))**2)/(2.*G))
24              R=QWC(N)
25              GO TO 30
26        20    IF(QWC(N).GT.QW(N)) GO TO 21
27              IF(QWC(N).LT.QW(N)) GO TO 22
28        21    YWU(N)=YWU(N)-0.005
29              GO TO 15
30              YWU(N)=YWU(N)+0.005
31              IF(YWU(N).GT.8.) GO TO 91
32              GO TO 15
33              I=I+1
34              K=N-I
35              QW(K)=QWC(K+1)
36              YWD(K)=YWU(K+1)
37              CL=CL+4.
38              WD(K)=(UW/RL)*CL+2.
39              CL=CL+4.
40              WU(K)=(UW/RL)*CL+2.
41        32    QU(K)=QW(K)+R
42              YWU(K)=B+0.005
43        35    B1=WU(K)*YWU(K)
44              IF(E-YWU(K).LT.0.001) GO TO 40
45              B2=(2.*G*(E-YWU(K)))**.5
46              B6=B1*B2
47              IF(ABS((QU(K)-B6)/QU(K)).LT.1.0E-1) GO TO 40
48              YWU(K)=YWU(K)=0.005
49              IF(YWU(K).GT.8.) GO TO 91
50              GO TO 35
51        40    YWA(K)=(YWD(K)+YWU(K))/2.
52              VU(K)=QU(K)/(WU(K)*YWU(K))
53              FU(K)=VU(K)/((G*YWU(K))**.5)
54              A1=3.*(FU(K)**2)
55              A2=(FU(K)**2)+2.
56              A3=A1/A2
57              CW=0.611*((1.-A3)**.5)
58              QWC(K)=0.67*WL*CW*((2.*G)**.5)*((YWA(K)-B)**1.5)
59              DQW=ABS(QWC(K)-QW(K))
60              IF(DQW.GT.0.01) GO TO 50
61              WRITE(6,31) QW(N),QWC(K),R,CW,YWU(K),YWD(K),FU(K),VU(K),K
62        31    FORMAT('0',5X,8F10.4,5X,I2)
63              GO TO 60
64        50    IF(QWC(K).LT.QW(K)) GO TO 51
65              IF(QWC(K).GT.QW(K)) GO TO 52
66        51    QW(K)=QW(K)-0.005
67              GO TO 32
68        52    QW(K)=QW(K)+0.005
69              IF(QW(K).GT.(2.*QD)) GO TO 93
70              GO TO 32
71        60    IF(K.EQ.1) GO TO 70
72              R=R+QWC(K)
73              GO TO 30
74        70    QT=QWC(1)+R
75              WRITE(6,71) QT
76        71    FORMAT('0',20X,F10.4)
77              DQ=ABS(QD-QT)
78              IF(QD.LT.(FLOAT(N)*0.05)) GO TO 80
```

```
79                     IF(QD.LT.QT) GO TO 75
80                     IF(QD.GT.QT) TO TO 76
81              75     QW(N)=QW(N)-0.01
82                     GO TO 10
83              76     QW(N)=QW(N)+0.05
84                     GO TO 10
85              80     DO 90 J=1,N,1
86                     WRITE(6,81) J,QWC(J),YWU(J)
87              81     FORMAT('0',1X,11HWEIR NUMBER,1X,I2,10X,11HFLOW EQUALS,
                       $1X,F10.4,10X,21HUPSTREAM DEPTH EQUALS,1X,F10.4)
88              90     CONTINUE
89                     GO TO 100
90              91     WRITE(6,92)
91              92     FORMAT('0',1X,17HYW GREATER THAN 8)
92                     GO TO 100
93              93     WRITE(6,94)
94              94     FORMAT('0',1X,19HQW GREATER THAN 2QD)
95             100     STOP
96                     END

            /GO
```

Computer Code for Example Problem 3-8

```
1                      DIMENSION QO(8),Y(8),FU(8),QU(8),CD(8),QOC(8),VU(8)
        C  QD=TOTAL INFLOW
        C  A=ORIFICE CROSS SECTIONAL AREA
        C  W=WIDTH OF CHANNEL
        C  WL=LENGTH OF EXIT WEIR
        C  N=NUMBER OF ORIFICES
        C  G=GRAVITY CONSTANT
        C  QP-FLOW THROUGH PROCESS UNIT
        C  NP=NUMBER OF PROCESS UNITS
        C  EL1=ELEVATION OF CHANNEL BOTTOM
        C  EL2=ELEVATION OF EXIT WEIR
2                      QD=23.21
3                      DO=.75
4                      A=(3.14*(DO**2))/4.
5                      W=4.0
6                      WL=24.0
7                      N=8
8                      G=32.2
9                      NP=2
10                     EL1=299.50
11                     EL2=300.00
12                     QP1=QD/FLOAT(NP)
13                     QP2=QP1
14                     QO(N)=QD/FLOAT(N)
15             10      H2=(QP2**.67)/((3.33**.67)*(WL**.67))
16                     H1=(QP1**.67)/((3.33**.67)*(WL**.67))
17                     Y(N)=0.01
18             20      VU(N)=QO(N)/(W*Y(N))
19                     FU(N)=VU(N)/((G*Y(N))**.5)
20                     CD(N)=0.611-(0.29*(FU(N)**2))
21                     DE=(EL1+Y(N)+((VU(N)**2)/(2.*G)))-(EL2+H2)
22                     IF(DE.LT.0.0) GO TO 25
23                     QOC(N)=CD(N)*A*((2.*G*DE)**.5)
24                     GO TO 26
25             25      Y(N)=Y(N)+0.01
26                     GO TO 20
27             26      DQO=ABS(QOC(N)-QO(N))
28                     IF(DQO.GT.0.01) GO TO 30
29                     I=0
30                     E=Y(N)+((VU(N)**2)/(2.*G))
31                     R=QOC(N)
```

```
32                   GO TO 40
33          30       IF(QOC(N).GT.QO(N)) GO TO 31
34                   IF(QOC(N).LT.QO(N)) GO TO 32
35          31       Y(N)=Y(N)-0.001
36                   GO TO 20
37          32       Y(N)=Y(N)+0.001
38                   IF(Y(N).GT.8.) GO TO 91
39                   GO TO 20
40          40       I=I+1
41                   K=N-I
42                   QO(K)=QOC(K+1)
43                   Y(K)=Y(K+1)
44          42       QU(K)=QO(K)+R
45          45       B1=W*Y(K)
46                   IF(E-Y(K).LT.0.001) GO TO 50
47                   B2=(2.*G*(E-Y(K)))**.5
48                   B6=B1*B2
49                   IF(ABS((QU(K)-B6)/QU(K)).LT.1.0E-1) GO TO 50
50                   IF(B6.GT.QU(K)) GO TO 46
51                   IF(B6.LT.QU(K)) GO TO 47
52          46       Y(K)=Y(K)+0.001
53                   GO TO 45
54          47       Y(K)=Y(K)-0.001
55                   GO TO 45
56          50       VU(K)=QU(K)/(W*Y(K))
57                   FU(K)=VU(K)/((G*Y(K))**.5)
58                   CD(K)=0.611-(0.29*(FU(K)**2))
C    THE NEXT CARD IS A CONTROL CARD
59                   IF(K.GT.4) GO TO 51
60                   DE=(ELI+Y(K)+((VU(K)**2)/(2.*G)))-(EL2+H1)
61                   GO TO 52
62          51       DE=(EL1+Y(K)+((VU(K)**2)/(2.*G)))-(EL2+H2)
63          52       IF(DE.LT.0.0) GO TO 62
64                   QOC(K)=CD(K)* *((2.*G*DE)**.5)
65                   DQW=ABS(QOC(K)-QO(K))
66                   IF(DQW.GT.0.01) GO TO 60
67                   WRITE(6,48) QO(N),QOC(K),R,CD(K),Y(K),FU(K),VU(K),K
68          48       FORMAT('0',5X,7F10.4,5X12)
69                   GO TO 70
70          60       IF(QOC(K).LT.QO(K)) GO TO 61
71                   IF(QOC(K).GT.QO(K)) GO TO 62
72          61       QO(K)=QO(K)-0.005
73                   GO TO 42
74          62       QO(K)=QO(K)+0.05
75                   GO TO 42
76          70       IF(K.EQ.1) GO TO 80
77                   R=R+QOC(K)
78                   GO TO 40
79          80       QT=QOC(1)+R
80                   WRITE(6,81) QT
81          81       FORMAT('0',20X,F10.4)
82                   DQ=ABS(QD-QT)
83                   IF(DQ.LT.(FLOAT(N)*0.01)) GO TO 95
84                   IF(QD.LT.QT) GO TO 85
85                   IF(QD.GT.QT) GO TO 86
C    THE NEXT TWO CARDS ARE CONTROL CARDS
86          85       QP2=QO(8)+QO(7)+QO(6)+QO(5)
87                   QP1=QO(4)+QO(3)+QO(2)+QO(1)
88                   QO(N)=QO(N)-0.005
89                   GO TO 10
C    THE NEXT TWO CARDS ARE CONTROL CARDS
90          86       QP2=QO(8)+QO(7)+QO(6)+QO(5)
91                   QP1=QO(4)+QO(3)+QO(2)+QO(1)
92                   QO(NL)=QO(N)+0.05
93                   GO TO 10
94          91       WRITE(6,92)
```

```
95              92    FORMAT('0',1X,16HY GREATER THAN 8)
96              95    STOP
97                    END

            /GO
```

Computer Code for Chapter 6: Distribution Channel for Primary Clarifier

```
 1                    DIMENSION QO(8),Y(8),FU(8),QU(8),CD(8),QOC(8),VU(8)
      C  QD=TOTAL INFLOW
      C  A=ORIFICE CROSS SECTIONAL AREA
      C  W=WIDTH OF CHANNEL
      C  WL=LENGTH OF EXIT WEIR
      C  N=NUMBER OF ORIFICES
      C  G=GRAVITY CONSTANT
      C  QP=FLOW THROUGH PROCESS UNIT
      C  NP=NUMBER OF PROCESS UNITS
      C  EL1=ELEVATION OF CHANNEL BOTTOM
      C  EL2=ELEVATION OF EXIT WEIR
 2                    QD=24.95/2.
 3                    WO=1.0
 4                    A=WO**2
 5                    W=4.0
 6                    WL=25.0
 7                    N=4
 8                    G=32.2
 9                    NP=2
10                    EL1=299.00
11                    EL2=300.50
12                    QP1=QD/FLOAT(NP)
13                    QP2=QP1
14                    QO(N)=QD/FLOAT(N)
15              10    H2=(QP2**.67)/((3.33**.67)*(WL**.67))
16                    H1=(QP1**.67)/((3.33**.67)*(WL**.67))
17                    Y(N)=0.01
18              20    VU(N)=QO(N)/(W*Y(N))
19                    FU(N)=VU(N)/((G*Y(N))**.5)
20                    CD(N)=0.611-(0.29*(FU(N)**2))
21                    DE=(EL1+Y(N)+((VU(N)**2)/(2.*G)))-(EL2+H2)
22                    IF (DE.LT.0.0) GO TO 25
23                    QOC(N)=CD(N)*A*((2.*G*DE)**.5)
24                    GO TO 26
25              25    Y(N)=Y(N)+0.01
26                    GO TO 20
27              26    DQO=ABS(QOC(N)-QO(N))
28                    IF(DQO.GT.0.01) GO TO 30
29                    I=0
30                    E=Y(N)+((VU(N)**2)/(2.*G))
31                    R=QOC(N)
32                    WRITE(6,49) QO(N),QOC(N),Y(N),FU(N),VU(N)
33              49    FORMAT('0',5X,5F10.4)
34                    GO TO 40
35              30    IF(QOC(N).GT.QO(N)) GO TO 31
36                    IF(QOC(N).LT.QO(N)) GO TO 32
37              31    Y(N)=Y(N)-0.001
38                    GO TO 20
39              32    Y(N)=Y(N)+0.001
40                    IF(Y(N).GT.8.) GO TO 91
41                    GO TO 20
42              40    I=I+1
43                    K=N-I
44                    QO(K)=QOC(K+1)
45                    Y(K)=Y(K+1)
46              42    QU(K)=QO(K)+R
```

```
47           45    B1=W*Y(K)
48                 IF(E-Y(K).LT.0.001) GO TO 50
49                 B2=(2.*G*(E-Y(K)))**.5
50                 B6=B1*B2
51                 IF(ABS((QU(K)-B6)/QU(K)).LT.1.0E-1) GO TO 50
52                 IF(B6.GT.QU(K)) GO TO 46
53                 IF(B6.LT.QU(K)) GO TO 47
54           46    Y(K)=Y(K)+0.001
55                 GO TO 45
56           47    Y(K)=Y(K)-0.001
57                 GO TO 45
58           50    VU(K)=QU(K)/(A*Y(K))
59                 FU(K)=VU(K)/((G*Y(K))**.5)
60                 CD(K)=0.611-(0.29*(FU(K)**2))
     C   THE NEXT CARD IS A CONTROL CARD
61                 IF(K.GT.2) GO TO 51
62                 DE=(EL1+Y(K)+((VU(K)**2)/(2.*G)))-(EL2+H1)
63                 GO TO 52
64           51    DE=(EL1+Y(K)+((VU(K)**2)/(2.*G)))-(EL2+H2)
65           52    IF(DE.LT.0.0) GO TO 62
66                 QOC(K)=CD(K)*A*((2.*G*DE)**.5)
67                 DQW=ABS(QOC(K)-QO(K))
68                 IF(DQW.GT.0.01) GO TO 60
69                 WRITE(6,48) QO(N),QOC(K),B,CD(K),Y(K),FU(K),VU(K),K
70           48    FORMAT('0',5X,7F10.4,5X.I2)
71                 GO TO 70
72           60    IF(QOC(K).LT.QO(K)) GO TO 61
73                 IF(QOC(K).GT.QO(K)) GO TO 62
74           61    QO(K)=QO(K)-0.005
75                 GO TO 42
76           62    QO(K)=QO(K)+0.05
77                 GO TO 42
78           70    IF(K.EQ.1) GO TO 80
79                 R=R+QOC(K)
80                 GO TO 40
81           80    QT=QOC(1)+R
82                 WRITE(6,81) QT
83           81    FORMAT('0',20X,F10.4)
84                 DQ=ABS(QD-QT)
85                 IF(DQ.LT.(FLOAT(N)*0.01)) GO TO 95
86                 IF(QD.LT.QT) GO TO 85
87                 IF(QD.GT.QT) GO TO 86
     C   THE NEXT TWO CARDS ARE CONTROL CARDS
88           85    QP2=QO(4)+QO(3)
89                 QP1=QO(2)+QO(1)
90                 QO(N)=QO(N)-0.005
91                 GO TO 10
     C   THE NEXT TWO CARDS ARE CONTROL CARDS
92           86    QP2=QO(4)+QO(3)
93                 QP1=QO(2)+QO(1)
94                 QO(N)=QO(N)+0.05
95                 GO TO 10
96           91    WRITE(6,92)
97           92    FORMAT('0',1X,16HY GREATER THAN 8)
98           95    STOP
99                 END

             /GO
```

Computer Code for Chapter 6: Distribution Channel for Aeration Basins

```
1                 DIMENSION QO(24),Y(24),FU(24),QU(24),CD(24),QOC(24),VU(24),
     C   QD=TOTAL INFLOW
     C   A=ORIFICE CROSS SECTIONAL AREA
```

```
      C    W=WIDTH OF CHANNEL
      C    WL=LENGTH OF EXIT WEIR
      C    N=NUMBER OF ORIFICES
      C    G=GRAVITY CONSTANT
      C    QP=FLOW THROUGH PROCESS UNIT
      C    NP=NUMBER OF PROCESS UNITS
      C    EL1=ELEVATION OF CHANNEL BOTTOM
      C    EL2=ELEVATION OF EXIT WEIR
2                  QD=24.9
3                  WO=0.5
4                  A=WO**2
5                  W=4.0
6                  WL=24.0
7                  N=24
8                  G=32.2
9                  NP=3
10                 EL1=294.8
11                 EL2=295.3
12                 QP1=QD/FLOAT(NP)
13                 QP2=QP1
14                 QP3=QP2
15                 QO(N)=QD/FLOAT(N)
16          10     H2=(QP2**.67)/((3.33**.67)*(WL**.67))
17                 H1=(QP1**.67)/((3.33**.67)*(WL**.67))
18                 H3=(QP3**.67)/((3.33**.67)*(WL**.67))
19                 Y(N)=0.01
20          20     VU(N)=QO(N)/(W*Y(N))
21                 FU(N)=VU(N)/((G*Y(N))**.5)
22                 CD(N)=0.611-(0.29*(FU(N)**2))
23                 DE=(EL1+Y(N)+((VU(N)**2)/(2.*G)))-(EL2+H2)
24                 IF(DE.LT.0.0) GO TO 25
25                 QOC(N)=CD(N)*A*((2.*G*DE)**.5)
26                 GO TO 26
27          25     Y(N)=Y(N)+0.01
28                 GO TO 20
29          26     DQO=ABS(QOC(N)-QO(N))
30                 IF(DQO.GT.0.01) GO TO 30
31                 I=0
32                 E=Y(N)+((VU(N)**2)/(2.*G))
33                 R=QOC(N)
34                 WRITE(6,49) QOC(N),Y(N),VU(N)
35          49     FORMAT('0',5X,3F10.4)
36                 GO TO 40
37          30     IF(QOC(N).GT.QO(N)) GO TO 31
38                 IF(QOC(N).LT.QO(N)) GO TO 32
39          31     Y(N)=Y(N)-0.001
40                 GO TO 20
41          32     Y(N)=Y(N)+0.001
42                 IF(Y(N).GT.8.) GO TO 91
43                 GO TO 20
44          40     I=I+1
45                 K=N-I
46                 QO(K)=QOC(K+1)
47                 Y(K)=Y(K+1)
48                 QU(K)=QO(K)+R
49          45     B1=W*Y(K)
50                 IF(E-Y(K).LT.0.001) GO TO 50
51                 B2=(2.*G*(E-Y(K)))**.5
52                 B6=B1*B2
53                 IF(ABS((QU(K)-B6)/QU(K)).LT.1.0E-1) GO TO 50
54                 IF(B6.GT.QU(K))  GO TO 46
55                 IF(B6.LT.QU(K)) GO TO 47
56          46     Y(K)=Y(K)+0.001
57                 GO TO 45
58          47     Y(K)=Y(K)-0.001
59                 GO TO 45
```

```
60          50    VU(K)=QO(K)/(W*Y(K))
61                FU(K)=VU(K)/((G*Y(K))**.5)
62                CD(K)=0.611-(0.29*(FU(K)**2))
        C   THE NEXT CARD IS A CONTROL CARD
63                IF(K.GT.16) GO TO 51
64                IF(K.GT.9) GO TO 53
65                DE=(EL1+Y(K)+((VU(K)**2)/(2.*G)))-(EL2+H2)
66                GO TO 52
67          51    DE=(EL1+Y(K)+((VU(K)**2)/(2.*G)))-(EL2+H3)
68                GO TO 52
69          53    DE=(EL1+Y(K)+((VU(K)**2)/(2.*G))))-(EL2+H1)
70          52    IF(DE.LT.0.0) GO TO 62
71                QOC(K)=CD(K)*A*((2.*G*DE)**.5)
72                DQW=ABS(QOC(K)-QO(K))
73                IF(DQW.GT.0.1) GO TO 60
74                WRITE(6,48) QO(N), QOC(K),R,CD(K),Y(K),FU(K),VU(K),K
75          48    FORMAT('0',5X,7F10.4,5X,I2)
76                GO TO 70
77          60    IF(QOC(K).LT.QO(K)) GO TO 61
78                IF(QOC(K).GT.QO(K)) GO TO 62
79          61    QO(K)=QO(K)-0.005
80                GO TO 42
81          62    QO(K)=QO(K)+0.05
82                GO TO 42
83          70    IF(K.EQ.1) GO TO 80
84                R=R+QOC(K)
85                GO TO 40
86          80    QT=QOC(1)+R
87                WRITE(6,81) QT
88          81    FORMAT('0',20X,F10.4)
89                DQ=ABS(QD-QT)
90                IF(DQ.LT.(FLOAT(N)*0.01)) GO TO 95
91                IF(QD.LT.QT) GO TO 85
92                IF(QD.GT.QT) GO TO 86
        C   THE NEXT TWO CARDS ARE CONTROL CARDS
93          85    QP3=QO(24)+QO(23)+QO(22)+QO(21)+QO(20)+QO(19)+
                  $QO(18)+QO(17)
94                QP2=QO(16)+QO(15)+QO(14)+QO(13)+QO(12)+QO(11)+QO(10)+
                  $QO(9)
95                QP1=QO(8)+QO(7)+QO(6)+QO(5)+QO(4)+QO(3)+QO(2)+QO(1)
96                QO(N)=QO(N)-0.005
97                GO TO 10
        C   THE NEXT TWO CARDS ARE CONTROL CARDS
98          86    QP3=QO(24)+QO(23)+QO(22)+QO(21)+QO(20)+QO(19)+QO(18)+
                  $QO(17)
99                QP2=QO(16)+QO(15)+QO(14)+QO(13)+QO(12)+QO(11)+QO(10)+
                  $QO(9)
100               QP1=QO(8)+QO(7)+QO(6)+QO(5)+QO(4)+QO(3)+QO(2)+QO(1)
101               QO(N)=QO(N)+0.05
102               GO TO 10
103         91    WRITE(6,92)
104         92    FORMAT('0',1X,16HY GREATER THAN 8)
105         95    STOP
106               END

            /GO
```

Computer Code for Chapter 6: Distribution Channel for Sludge Return

```
1             DIMENSION QO(8),Y(8),FU(8),QU(8),CD(8),QOC(8),VU(8)
        C   QD=TOTAL INFLOW
        C   A=ORIFICE CROSS SECTIONAL AREA
        C   W=WIDTH OF CHANNEL
        C   WL=LENGTH OF EXIT WEIR
```

```
        C   N=NUMBER OF ORIFICES
        C   G=GRAVITY CONSTANT
        C   QP=FLOW THROUGH PROCESS UNIT
        C   NP=NUMBER OF PROCESS UNITS
        C   EL1=ELEVATION OF CHANNEL BOTTOM
        C   EL2=ELEVATION OF EXIT WEIR
 2                      QD=13.7
 3                      WO=0.75
 4                      A=WO**2
 5                      W=4.0
 6                      WL=25.0
 7                      N=6
 8                      G=32.2
 9                      NP=3
10                      EL1=294.8
11                      EL2=295.3
12                      QP1=QD/FLOAT(NP)
13                      QP2=QP1
14                      QP3=QP2
15                      QO(N)=QD/FLOAT(N)
16          10          H2=(QP2**.67)/((3.33**.67)*(WL**.67))
17                      H1=(QP1**.67)/((3.33**.67)*(WL**.67))
18                      H3=(QP3**.67)/((3.33**.67)*(WL**.67))
19                      Y(N)=0.01
20          20          VU(N)=QO(N)/(W*Y(N))
21                      FU(N)=VU(N)/((G*Y(N))**.5)
22                      CD(N)=0.611-(0.29*(FU(N)**2))
23                      DE=(EL1+Y(N)+((VU(N)**2)/(2.*G)))-(EL2+H2)
24                      IF(DE.LT.0.0) GO TO 25
25                      QOC(N)=CD(N)*A*((2.*G*DE)**.5)
26                      GO TO 26
27          25          Y(N)=Y(N)+0.01
28                      GO TO 20
29          26          DQO=ABS(QOC(N)-QO(N))
30                      IF(DQO.GT.0.05) GO TO 30
31                      I=0
32                      E=Y(N)+((VU(N)**2)/(2.*G))
33                      R=QOC(N)
34                      WRITE(6,49) QO(N),QOC(N),Y(N),FU(N),VU(N)
35          49          FORMAT('0',5X,5F10.4)
36                      GO TO 40
37          30          IF(QOC(N).GT.QO(N)) GO TO 31
38                      IF(QOC(N).LT.QO(N)) GO TO 32
39          31          Y(N)=Y(N)-0.001
40                      GO TO 20
41          32          Y(N)=Y(N)+0.001
42                      IF(Y(N).GT.8.) GO TO 91
43                      GO TO 20
44          40          I=I+1
45                      K=N-I
46                      QO(K)=QOC(K+1)
47                      Y(K)=Y(K+1)
48          42          QU(K)=QO(K)+R
49          45          B1=W*Y(K)
50                      IF(E-Y(K).LT.0.001) GO TO 50
51                      B2=(2.*G*(E-Y(K)))**.5
52                      B6=B1*B2
53                      IF(ABS((QU(K)-B6)/QU(K)).LT.1.0E-1) GO TO 50
54                      IF(B6.GT.QU(K)) GO TO 46
55                      IF(B6.LT.QU(K)) GO TO 47
56          46          Y(K)=Y(K)+0.001
57                      GO TO 45
58          47          Y(K)=Y(K)-0.001
59                      GO TO 45
60          50          VU(K)=QU(K)/(W*Y(K))
```

```
61              FU(K)=VU(K)/((G*Y(K))**.5)
62              CD(K)=0.611-(0.29*(FU(K)**2))
     C   THE NEXT CARD IS A CONTROL CARD
63              IF(K.GT.4) GO TO 51
64              IF(K.LT.3) TO TO 53
65              DE=(EL1+Y(K)+((VU(K)**2)/(2.*G)))-(EL2+H2)
66              GO TO 52
67         51   DE=(EL1+Y(K)+((VU(K)**2)/(2.*G)))-(EL2+H3)
68              GO TO 52
69         53   DE=(EL1+Y(K)+((VU(K)**2)/(2.*G)))-(EL2+H1)
70         52   IF(DE.LT.0.0) GO TO 62
71              QOC(K)=OD(K)*A*((2.*G*DE)**.5)
72              DQW=ABS(QOC(K)-QO(K))
73              IF(DQW.GT.0.01) GO TO 60
74              WRITE(6,48) QO(N),QOC(K),R,CD(K),Y(K),FU(K),VU(K),K
75         48   FORMAT('0',5X,7F10.4,5X,I2)
76              GO TO 70
77         60   IF(QOC(K).LT.QO(K)) GO TO 61
78              IF(QOC(K).GT.QO(K)) GO TO 62
79         61   QO(K)=QO(K)-0.005
80              GO TO 42
81         62   QO(K)=QO(K)+0.05
82              GO TO 42
83         70   IF(K.EQ.1) GO TO 80
84              R=R+QOC(K)
85              GO TO 40
86         80   QT=QOC(1)+R
87              WRITE(6,81) QT
88         81   FORMAT('0',20X,F10.4)
89              DQ=ABS(QD-QT)
90              IF(DQ.LT.(FLOAT(N)*0.01)) GO TO 95
91              IF(QD.LT.QT) GO TO 85
92              IF(QD.GT.QT) GO TO 86
     C   THE NEXT TWO CARDS ARE CONTROL CARDS
93         85   QP3=QO(6)+QO(5)
94              QP2=QO(4)+QO(3)
95              QP1=QO(2)+QO(1)
96              QO(N)=QO(N)-0.005
97              GO TO 10
     C   THE NEXT TWO CARDS ARE CONTROL CARDS
98         86   QP3=QO(6)+QO(5)
99              QP2=QO(4)+QO(3)
100             QP1=QO(2)+QO(1)
101             QO(N)=QO(N)+0.05
102             GO TO 10
103        91   WRITE(6,92)
104        92   FORMAT('0',1X,16HY GREATER THAN 8)
105        95   STOP
106             END

          /GO
```

APPENDIX II: PROPERTIES OF WATER

Temperature ($^\circ F$)	Specific weight, γ (lb/ft^3)	Mass density, ρ ($lb\text{-}sec^2/ft^4$)	Dynamic viscosity, $\mu \times 10^5$ ($lb\text{-}sec/ft^2$)	Kinematic viscosity, $\nu \times 10^5$ (ft^2/sec)	Vapor pressure head P_V/γ (ft)
32	62.42	1.940	3.746	1.931	0.20
40	62.43	1.938	3.229	1.664	0.28
50	62.41	1.936	2.735	1.410	0.41
60	62.37	1.934	2.359	1.217	0.59
70	62.30	1.931	2.050	1.059	0.84
80	62.22	1.927	1.799	0.930	1.17
90	62.11	1.923	1.595	0.826	1.61
100	62.00	1.918	1.424	0.739	2.19
110	61.86	1.913	1.284	0.667	2.95
120	61.71	1.908	1.168	0.609	3.91
130	61.55	1.902	1.069	0.558	5.13
140	61.38	1.896	0.981	0.514	6.67
150	61.20	1.890	0.905	0.476	8.58
160	61.00	1.896	0.838	0.442	10.95
170	60.80	1.890	0.780	0.413	13.83
180	60.58	1.883	0.726	0.385	17.33
190	60.36	1.876	0.678	0.362	21.55
200	60.12	1.868	0.637	0.341	26.59
212	59.83	1.860	0.593	0.319	33.90

APPENDIX III: PROGRAMMABLE CALCULATOR PROGRAMS FOR SELECTED PROBLEMS USING AN HP41C AND PRINTER

Program for Example Problem 1-9: The iterative procedure described and illustrated in Chap. 1 for divided-flow manifolds can also be carried out using the following equations:

$$\beta_i = \phi(a/q_iA)^2\left(\sum_{k=i}^{N} q_k\right)^2 + \Theta + 1.0 \qquad \text{(A-1)}$$

$$q_1 = (Q_0/\sqrt{\beta_1})\left(\sum_{i=1}^{N} \sqrt{1/\beta_i}\right)^{-1} \qquad \text{(A-2)}$$

$$q_i = q_1\sqrt{\beta_1}\sqrt{1/\beta_i} \qquad \text{(A-3)}$$

These equations were derived from the equations in Sec. 1.5 by noting that

$$V_{N_i} = \frac{1}{A}\sum_{k=i}^{N} q_k$$

By initally assuming $q = q_i = Q_0/N$ for each lateral, these three equations can be solved iteratively (in order) until the desired accuracy in q's is obtained.

```
01♦LBL "MANIFLD"
FIX 3 "NO. LATERALS?"
PROMPT STO 00
"LAT. AREA?" PROMPT
STO 01 "MAN. AREA?"
PROMPT STO 02
"TOTAL FLOW?" PROMPT
STO 03 "PHI?" PROMPT
STO 04 "THETA?" PROMPT
STO 05 RCL 00 1000 /
1 + STO 07 STO 08
RCL 01 RCL 02 /
STO 06 RCL 03 RCL 00
/ STO 09

36♦LBL A
RCL 07 INT 20 +
STO 10 RCL 09
STO IND 10 ISG 07
GTO A

46♦LBL "REPEAT"
RCL 08 STO 07

49♦LBL B
RCL 07 INT STO 10 20
+ STO 14 RCL 00 1000
/ RCL 10 + STO 11 0
STO 13

64♦LBL C
RCL 11 INT 20 +
STO 12 RCL IND 12
RCL 13 + STO 13
ISG 11 GTO C RCL 10
RCL 00 + 20 + STO 15
RCL 04 SQRT RCL 06 *
```

```
RCL IND 14 / RCL 13 *
X↑2 RCL 05 + 1 +
STO IND 15 ISG 07
GTO B RCL 00 21 +
STO 16 RCL 08 STO 07
0 STO 17 ADV CLA
"LATERAL DISCHA" ACA
"RGE" ACA PRBUF CLA
" NO.    IN CF" ACA
"S" ACA PRBUF ADV

120♦LBL D
RCL 07 INT 20 +
RCL 00 + STO 14
RCL IND 14 1/X SQRT
RCL 17 + STO 17
ISG 07 GTO D 1/X
RCL 03 * RCL IND 16
SQRT / STO 21
"    1.    " ARCL X
PRA RCL 00 1000 / 2
+ STO 19

152♦LBL E
CLA " " ACA FIX 0
RCL 19 INT ACX 20 +
STO 09 CLA "     "
ACA FIX 3 RCL 00 +
STO 15 RCL IND 15 1/X
SQRT RCL 21 *
RCL IND 16 SQRT *
STO IND 09 ACX PRBUF
ISG 19 GTO E 0
"REPEAT?" PROMPT X≠0?
GTO "REPEAT" STOP
.END.
```

LATERAL NO.	DISCHARGE IN CFS	LATERAL NO.	DISCHARGE IN CFS
1.	1.638	1.	1.492
2.	1.729	2.	1.649
3.	1.821	3.	1.796
4.	1.912	4.	1.928
5.	1.997	5.	2.041
6.	2.072	6.	2.131
7.	2.132	7.	2.197
8.	2.169	8.	2.237

LATERAL NO.	DISCHARGE IN CFS	LATERAL NO.	DISCHARGE IN CFS
1.	1.546	1.	1.484
2.	1.673	2.	1.646
3.	1.800	3.	1.797
4.	1.919	4.	1.930
5.	2.025	5.	2.043
6.	2.113	6.	2.133
7.	2.177	7.	2.199
8.	2.217	8.	2.239

LATERAL NO.	DISCHARGE IN CFS	LATERAL NO.	DISCHARGE IN CFS
1.	1.508	1.	1.480
2.	1.655	2.	1.645
3.	1.796	3.	1.797
4.	1.925	4.	1.931
5.	2.036	5.	2.044
6.	2.126	6.	2.134
7.	2.191	7.	2.200
8.	2.231	8.	2.240

Program for Example Problem 2-1: The output for the final design of Example Problem 2-1 is given below. These results correspond to Columns 1, 4, and 5 of Table 2-5. It should be noted that the program that is presented can also be used to solve Example Problems 2-2 and 2-3.

```
01♦LBL "DIFUSR"
"GAMMA R?" PROMPT
STO 14 "GAMMA E?"
PROMPT STO 13 RCL 14
- X=0? GTO C "SLOPE?"
PROMPT STO 15

15♦LBL C
"E1?" PROMPT

18♦LBL B
STO 08 0 STO 10
STO 16 STO 23 STO 24
"NO. SCTS?" PROMPT
STO 00 1000 / 1 +
STO 18

33♦LBL D
```

```
RCL 08 / .58 * CHS
.63 + PI * 4 /
RCL 03 X↑2 * RCL 08
64.4 * SQRT * STO 09
RCL 21 + STO 21
RCL 09 RCL 16 +
STO 16 4 * PI /
RCL 04 X↑2 / STO 10
X↑2 RCL 06 * RCL 05
* 64.4 / RCL 04 /
RCL 08 STO 22 +
RCL 14 RCL 13 -
RCL 13 / RCL 15 *
RCL 05 * + STO 08
FIX 0 9 RCL 23 X≠Y?
XEQ 02 ACX " " ACA
FIX 3 10 RCL 09 X≠Y?
XEQ 02 ACX " " ACA
```

SECTION 1.

PORT NO.	PORT Q	PIPE V
1.	1.774	0.361
2.	1.776	0.723
3.	1.775	1.085
4.	1.771	1.446
5.	1.766	1.805
6.	1.758	2.163
7.	1.748	2.520
8.	1.737	2.873
9.	1.724	3.225
10.	1.710	3.573

SECTION Q = 17.540

```
RCL 24  1  +  STO 24  0        10  RCL 10  X<=Y?
STO 20  STO 21                 XEQ 02  ACX  "  "  ACA
"PRTS IN SCN?"  PROMPT         PRBUF  FIX 3  RCL 20
STO 02  1000  /  1  +          RCL 02  -  X=0?  XEQ 01
STO 19  "PORT DIA?"            ISG 19  GTO E  ISG 18
PROMPT  STO 03                 GTO D  ADV  RCL 16  CLA
"SPACING?"  PROMPT             "TOTAL Q = "  ARCL X
STO 05  "PIPE DIA?"            PRA  ADV  RCL 22  CLA
PROMPT  STO 04                 "EN = "  ARCL X  PRA
"F-VALUE?"  PROMPT             ADV  0  "NEW E1?"
STO 06  FIX 0  RCL 24          PROMPT  X≠0?  GTO B
"  SECTION "  ARCL X           STOP
PRA  ADV  FIX 3  CLA
"PORT    PORT "  ACA           205♦LBL 01
"  PIPE"  ACA  PRBUF           RCL 21  ADV  CLA
CLA  " NO.      Q    "         "SECTION Q = "  ARCL X
ACA  "   V"  ACA  PRBUF        PRA  ADV  RTN
ADV
                              214♦LBL 02
81♦LBL E                       "  "  ACA  RTN  END
RCL 23  1  +  STO 23
RCL 20  1  +  STO 20
RCL 10  X↑2  64.4  /
```

```
              SECTION 2.

PORT     PORT      PIPE
NO.       Q         V

11.     1.829     3.946
12.     1.810     4.314
13.     1.791     4.673
14.     1.772     5.040
15.     1.753     5.397
16.     1.734     5.750
17.     1.717     6.100
18.     1.700     6.446
19.     1.685     6.790
20.     1.671     7.130

SECTION Q = 17.460

TOTAL Q = 35.000

EN = 2.939
```

Program for Example Problem 3-4:   The equations in the lateral inflow section can be combined to give

$$y_2 = y_1 + S(x_2 - x_1) - S_E(x_2 - x_1) - \frac{Q_1(V_2 + V_1)}{g(Q_2 + Q_1)}\left[(V_2 - V_1) + \frac{V_2}{Q_1}(Q_2 - Q_1)\right]$$

This is solved iteratively in the program given below. The output from the program for Example Problem 3-4 is presented. Convergence of y_2 within 0.001 ft. was required.

```
01♦LBL "LATINFL"              105♦LBL "Y2-CALC"
FIX 3  "WIDTH?"  PROMPT       RCL 16  RCL 17  +  X↑2
STO 00  "LENGTH?"             RCL 01  X↑2  *  8.83  /
PROMPT  STO 06  "N?"          RCL 14  1.3333  Y↑X  /
PROMPT  STO 01                STO 21  RCL 08  RCL 07
"BED SLOPE?"  PROMPT          -  RCL 17  *  RCL 07  /
STO 02  "TOTAL Q?"            RCL 17  RCL 16  -  +
PROMPT  STO 07                RCL 07  *  32.2  /
"DELTA X?"  PROMPT            RCL 17  RCL 16  +  *
STO 04  "D.S. DEPTH?"         RCL 07  RCL 08  +  /
PROMPT  STO 05  RCL 07        CHS  RCL 05  +  RCL 02
RCL 06  /  STO 03             RCL 15  RCL 06  -  *  +
RCL 05  STO 18  ADV  CLA      RCL 21  RCL 15  RCL 06
"  SECTION    "  ACA          -  *  -  STO 22  RCL 18
" DEPTH"  ACA  PRBUF          -  1000  *  ABS  1  X>Y?
ADV  CLA  "    "  ACA         GTO D  RCL 22  STO 18
10  FIX 1  RCL 06  X<Y?       GTO C
XEQ 01  ACX  "    "
ACA  10  FIX 3  RCL 05        170♦LBL D
X<Y?  XEQ 01  ACX  PRBUF      FIX 1  CLA  "    "  ACA
                             10  RCL 15  X<Y?  XEQ 01
56♦LBL A                      ACX  "    "  ACA  FIX 3
RCL 05  RCL 00  *             10  RCL 22  X<Y?  XEQ 01
STO 09  RCL 05  2  *          ACX  PRBUF  RCL 22
RCL 00  +  /  STO 11          STO 05  RCL 08  STO 07
RCL 07  RCL 09  /             RCL 15  STO 06  RCL 12
STO 16                        STO 11  RCL 17  STO 16
                             RCL 15  RCL 04  X<Y?
```

```
SECTION      DEPTH

24.0        0.801
23.0        0.992
22.0        1.062
21.0        1.112
20.0        1.152
19.0        1.186
18.0        1.214
17.0        1.239
16.0        1.260
15.0        1.280
14.0        1.296
13.0        1.311
12.0        1.324
11.0        1.336
10.0        1.346
9.0         1.355
8.0         1.363
7.0         1.369
6.0         1.375
5.0         1.379
4.0         1.382
3.0         1.385
2.0         1.386
1.0         1.386
```

```
 72◆LBL B                    GTO B
RCL 06  RCL 04  -
STO 15  RCL 07  RCL 03     203◆LBL 01
RCL 04  *  -  STO 08        " "  ACA  RTN  STOP  END

 83◆LBL C
RCL 18  RCL 00  *
STO 10  RCL 18  2  *
RCL 00  +  /  STO 12
RCL 08  RCL 10  /
STO 17  RCL 11  RCL 12
+  2  /  STO 14
```

Program for Example Problem 3-6: In treatment plants, the depth at the downstream end of the side weir will usually be set by a downstream control rather than by friction. Consequently, the program prompts for inputs of the weir height and for the depth at the downstream end of the weir for the design flow, rather than calculates them according to the Manning equation.

The output for inputs of $W = 6$ ft, Q design $= 30$ cfs, Q weir $= 5$ cfs, $B = 1.128$ ft, and $y_{ds} = 1.315$ ft is shown below together with the calculator program.

```
 01◆LBL "SD WEIR"           99◆LBL "CONST"
RAD  "WIDTH?"  PROMPT      RCL 20  RCL 21  -
STO 00  "QDESIGN?"         RCL 20  RCL 24  -  /
PROMPT  STO 04  "QWEIR?"   SQRT  ASIN  3  *  STO 25
PROMPT  STO 05             RCL 20  RCL 21  -
"WEIR HT B?"  PROMPT       RCL 21  RCL 24  -  /
STO 24  "DS DESIGN Y?"     SQRT  RCL 20  2  *
PROMPT  STO 18             RCL 24  3  *  -  *
                          RCL 20  RCL 24  -  /
 18◆LBL "SP EN"            CHS  RCL 25  +  STO 26
RCL 04  RCL 05  -          ADV  CLA
RCL 18  /  RCL 00  /       "INTGR CONST = "  ACA
X↑2  64.4  /  RCL 18  +    RCL 26  ACX  " FT."  ACA
STO 20  ADV  CLA           PRBUF
"DESIGN E =    "  ACA
RCL 20  ACX  " FT."  ACA
PRBUF                     145◆LBL "WEIR L"
                          RCL 20  RCL 18  -
 41◆LBL "Y US"            RCL 20  RCL 24  -  /
RCL 04  X↑2  64.4  /       SQRT  ASIN  3  *  STO 25
RCL 00  X↑2  /  STO 22     RCL 20  RCL 18  -
RCL 18  XEQ 01  STO 17     RCL 18  RCL 24  -  /
RCL 18  STO 07  1.1  *     SQRT  RCL 20  2  *
STO 06  XEQ 01  STO 16     RCL 24  3  *  -  *
                          RCL 20  RCL 24  -  /
 60◆LBL A                  RCL 25  -  RCL 26  +
RCL 07  RCL 06  -          RCL 00  *  4.1  /  64.4
RCL 17  *  RCL 17          SQRT  *  STO 27  ADV
RCL 16  -  /  CHS          CLA  "WEIR LENGTH ="
RCL 07  +  STO 21          ACA  RCL 27  ACX  " FT."
RCL 07  STO 06  RCL 17     ACA  PRBUF
STO 16  RCL 21  STO 07
XEQ 01  STO 17  RCL 16     199◆LBL 01
-  1000  *  ABS  1        STO 23  X↑2  1/X  RCL 22
X<=Y?  GTO A  ADV  CLA     *  RCL 23  +  RCL 20  -
"US DESIGN Y = "  ACA      RTN  END
RCL 21  ACX  " FT."  ACA
PRBUF
```

```
DESIGN E =      1.471 FT.

US DESIGN Y =   1.202 FT.

INTGR CONST =   5.709 FT.

WEIR LENGTH = 27.141 FT.
```

INDEX